The Aztec Fascination with Birds:
Deciphering 16th-Century Sources in Náhuatl

CONTRIBUTIONS
IN ETHNOBIOLOGY

CONTRIBUTIONS IN ETHNOBIOLOGY

Cynthia Fowler and Steve Wolverton, Series Editors
Cheryl Takahashi, Managing Editor

Contributions in Ethnobiology is a peer-reviewed monograph series presenting original book-length data-rich, state-of-the-art research in ethnobiology. It is the only monograph series devoted expressly to representing the breadth of ethnobiological topics.

The Aztec Fascination with Birds: Deciphering 16th-Century Sources in Náhuatl

Eugene S. Hunn

Society of Ethnobiology

2023

Library of Congress Control Number: 2023905766

ISBN 978-0-9990759-8-2 (paperback)
ISBN 978-0-9990759-9-9 (PDF)

Society of Ethnobiology
Boston University Archaeology Room 345,
675 Commonwealth Ave., Boston, MA 02215

Cover: Top: HUĪTZITZI-LI, hummingbirds (Dibble & Anderson, 1963, Figure 61, after Paso y Troncoso); Bottom left: TLĀUH-QUECHŌL, the Roseate Spoonbill (Dibble & Anderson, 1963, Figure 44, after Paso y Troncoso); Bottom right: CUĀUH-TLI$_2$, Golden Eagle (Dibble & Anderson, 1963, Figure 114, after Paso y Troncoso).

Table of Contents

Eugene S. Hunn

List of Figures

List of Sound Clips

- Russet-naped Wood-Rail, sound clip by Jessy Lopez Herra, ML435824171
- Thicket Tinamou, sound clip by Vicente Desjardins, ML435021671
- Great Tinamou, sound clip by Adam Dudley, ML432084251
- Aztec Rail, sound clip by Anuar López, ML432093161
- Belted Kingfisher, sound clip by Liza Verkony, ML436910561
- Laughing Falcon, sound clip by Adam Betuel, ML436679641
- Green Shrike-Vireo, sound clip by Daniel Garrigues, ML 437064121
- Canyon Wren, sound clip by Brayden Luikart, ML436865601
- Curve-billed Thrasher, sound clip by Andrew Theus, ML436112051

Sound clips along with images are archived in The Aztec Fascination with Birds' Digital Collection of Images and Sounds. Scan the QR code below to access the archive, or visit it at: https://ethnobiology.org/publications/contributions/aztec-fascination-birds/ digital-collection.

Acknowledgments

This project would not have gotten off the ground had I not cold-called Stephanie Wood, editor of the Online Náhuatl Dictionary housed at the University of Oregon. Dr. Wood enthused about the notion of revisiting the *Florentine Codex* birds as a means to sharpen the translations of the Náhuatl avian nomenclature. Her advice and encouragement steered me away from potential misreadings of the Náhuatl texts. I have also benefited from insightful comments by Náhuatl scholar, León García Garagarza. I would like to credit also Cynthia Fowler, Cheryl Takahashi, and Steve Wolverton for their timely editing, and the Society of Ethnobiology editorial staff of *Contributions in Ethnobiology* for making the rich color printing possible and thank two anonymous reviewers for support and for valuable critical suggestions.

Permissions to reproduce extended text quotes, images, and sound clips in the print version and the accompanying digital collection are as follows:

- Quoted material from the *Florentine Codex* (FC): Reprinted from *Florentine Codex: General History of the Things of New Spain, Book 11, Earthly Things¸* by Fray Bernardino de Sahagún, translated by Arthur J. O. Anderson, and Charles E. Dibble, with permission from School for American Research, Santa Fe, New Mexico. Copyright 1981.
- The Macaulay Library at the Cornell Lab of Ornithology for permission to reproduce the photographs and sound clips of the birds described (Figures 8, 9, 10, 12, 13, 14, 15, 16, 17, 18, 19, 20, 21, 23, 24, 26, 27, 28, 29, 30, 33, 35, and 38 and all sound clips). Each image and sound clip, identified by the Macaulay Library catalog number, is credited to the photographer and/or sound recorder in the respective figure captions.
- The Kunsthistorisches Museum, Wien, for permission to reproduce the image and description of "Moteuczoma's headdress" (Figure 6).
- The Fine Arts Museums of San Francisco for permission to reproduce the color image entitled "Small bird with shield, spear, and flower. Techinantitla, Teotihuacan, AD 600-750," from *Feathered Serpents and Flowering Trees: Reconstructing the Murals of Teotihuacan* (Berrin, 1988, p. 170, Plate 17) (Figure 5).
- The American Ornithological Society for permission to reproduce in color the painting by Andrew Jackson Grayson published in the *Distributional Checklist of the Birds of Mexico, Part II*, Miller et al., 1957, pp. 62–63 (Figure 25).
- The Akademische Druck- und Verlagsanstalt for permission to reproduce the image of Quetzalcoatl from the *Codex Borbonicus* (Figure 7).
- To Symbolikon for the graphic image of Ehecatl, the wind god (Figure 36).

Imperial Aztec Ethno-Ornithology: Introduction

Imagine you are living in the Valley of Mexico; it is 1519; mysterious floating mountains—a first impression of the Spanish ships—have just been spotted off the eastern coast. Strange weather phenomena disturb the placid surface of the lakes that fill the Valley as the Water People in their canoes hunt ducks for dinner:

> The wind lashed the water until it boiled. It was as if it were boiling with rage, as if it were shattering itself in its frenzy. It began from far off, rose high in the air and dashed against the walls of the houses. The flooded houses collapsed into the water (quoted in Leon-Portilla, 1992, p. 5).

With some foreboding these Aztec commoners feared a cataclysm awaited. Could they have imagined that their great capital city, Tenochtitlán, commanding an island in the center of the lakes (Figure 1), would in just two years' time be destroyed by an invading force of bearded men and by a terrible plague? By 1521 their shining city, "the Mexican Venice," is in ruins, their *Tlatoani*, the emperor Mocteuczoma Xocoyotzin (Figure 2), assassinated, their temple pyramids torn down, their people decimated. But daily life in nearby rural precincts survived this apocalypse.

Six years pass and brown robed men arrive to preach a new religion, intent on erasing all memory of the ancient Mesoamerican gods and the bloody sacrifices their gods had required to "feed the sun." One of these missionaries from across the ocean was Fray Bernardino de Sahagún (Figure 3). He believed that to truly convert the children of the Aztec elders required first of all that the missionary monks fully understand the old ways and the alien beliefs of their potential converts. To that end a college was established to train the children of the Aztec nobility to record their world, writing with a Latin alphabet but in their own words, in their own language, Náhuatl, an encyclopedia of Aztec knowledge, of the gods, the rulers, the common people, their history, ending with an account of the "Earthly Things" of their world, including, with the rest, "all the birds." My task here is to attempt to translate as best I can at this late date the names of these birds, so well-known to the Aztecs, in order to better appreciate the profound appreciation of the Aztec people for their bird neighbors.

How this encyclopedia of Nahua culture survived centuries of harrowing misadventures is an epic tale. The hand-written draft that survives was edited by the 80-year-old Sahagún in the midst of an epidemic of hemorrhagic fever.

Figure 1. The Valley of Mexico circa 1519 showing settlement names and water systems (Madman, 2001 CC BY-SA 3.0).

Figure 2. Moteuczoma Xocoyotzin (a.k.a. Montezuma II), painted by Antonio Rodríguez.

Figure 3. Fray Bernardino de Sahagún, Museo Nacional de Historia ().

Writing during the virulent epidemic of 1576, Sahagún recalled an earlier plague of 1545, when he claimed to have buried more than ten thousand corpses before he himself fell ill and nearly died. 'There is hardly anyone still in the college,' he lamented. 'Dead and sick, almost all gone' (quoted in Terraciano, 2019, location 105–106).

"[B]y the 1570s, Sahagún had seen the 'whole land ... depleted of people' with his very own eyes" (quoted in Terraciano, 2019, location 114).

Meanwhile, back in Spain, in April of 1577, King Philip II issued a decree that concluded: "You will be advised not to permit anyone, for any reason, in any language, to write concerning the superstitions and way of life these Indians had" (quoted from Códice Franciscano in Terraciano, 2019, p. 8). Apparently in defiance of this royal order, "Sahagun entrusted a complete version with images to his Franciscan colleague, [Fray Rodrigo de] Sequera, who departed Mexico for Spain in 1580" (quoted from Códice Franciscano in Terraciano, 2019, p. 8). At this point, the trail goes cold. However, we know that sometime before 1588 Ferdinando I de' Medici acquired the manuscript, by gift or purchase, from Sequera (Terraciano, 2019, p. 8). It was then squirreled away in an archive in Florence, now in Italy. Hence the title *Florentine Codex*. Here it was "lost" for 300 years. Today, photocopies of the original manuscript—hand-written and hand painted in glorious color—may be viewed on-line while a printed version is readily available which features an English translation matched paragraph-by-paragraph to the Náhuatl original (Dibble & Anderson, 1957–1982). We now may slip through this time-warp to hear the words of the Aztec scribes in "real time."

A brief note on spelling conventions might assist the reader here. The Náhuatl in the original manuscript used "x" for a sound most closely approximated in modern English as "sh." So "Mexico" as now pronounced evolved by stages from the original Náhuatl MĒXIHCA-TL, that is, "resident of Mexico-Tenochtitlán" (Karttunen, 1983, p. 145) pronounced ~ "meshica." I use all caps for "canonical forms" of Náhuatl terms, that is, forms verified linguistically and represented as such in modern Náhuatl dictionaries. Classical Náhuatl distinguished long from short vowels by superimposing a horizontal bar over the long vowel, and the "H" represents a glottal stop, roughly approximated by the throaty sound of English "h," as in "heart." I also hyphenate these canonical forms to indicate the internal structure of the word, isolating suffixes and isolating morphemes. For example, QUETZAL-TŌTŌ-TL is compounded of QUETZAL, a precious feather, TŌTŌ-, "bird," and -TL, marking the singular. The Aztec scribes wrote "j" as a vowel equivalent to the "i" sound of modern Spanish and used the cedilla, "ç," for "z." When I quote the *Codex*, I may use these symbols. The "tl" sound is common in Náhuatl and sounds just like the "t-l" we hear in "at last," but a combination that serves as an independent sound.

Who Were the Aztecs?

According to legend, the Aztec people were poor immigrants from a near-mythical place of origin northwest of the Valley of Mexico known as *Aztlan*, which might be glossed as "the place of egrets." On their arrival ca. AD 1300, they found the lake shores of the Valley of Mexico densely settled and controlled by several powerful city-states. All harked back to a predecessor Toltec civilization with its capital, Tula, by then a ruin. The legendary account has these poor migrants relegated to a swampy island in Lake Texcoco, reclaimed where they found an eagle perched on a cactus clutching a snake (Figure 4). These Aztecs were first subjects of Colhuacan, the power to the south, then allies of the Tepanec kingdom of Atzcapotzalco, which controlled the western shore of the lake. By 1426 the Aztecs turned the tables on their powerful allies, defeating Atzcapotzalco to become the dominant force in the Valley of Mexico (cf. Leon-Portilla, 1992).

Figure 4. Founding of Tenochtitlán from the *Codex Mendoza*.

This conflict-ridden legendary history distorts a fundamental fact, that the Aztecs were but one of many local communities sharing a long and rich cultural heritage dating at least to the great city Teotihuacán, northeast of present-day Mexico City, heart of the first of three successive Valley of Mexico empires, based at Teotihuacán, Tula, and Tenochtitlán. Teotihuacán was a true city, well organized with an estimated urban population of nearly 200,000 at its heyday ca. AD 500. Its reach extended as far as Guatemala. The language spoken in that city is still not known but was likely a central Mexican language other than Náhuatl. What I characterize here as "the Aztec fascination with birds" may be traced directly to that first great central Mexican civilization based at Teotihuacán. This is apparent from the recent discovery of polychrome murals covering the walls of a group of basement rooms near the Pyramid of the Moon at the eastern end of the broad "Avenue of the Dead," the central axis of the city (Figure 5).

Figure 5. Small bird with shield, spear, and flower. Techinantitla, Teotihuacan, AD 600–750 (Berrin, 1988, p. 170, Plate 17; with permission of The Fine Arts Museums of San Francisco).

Birds are among the most frequent subjects represented in Teotihuacan art…. Despite their ubiquity and importance, the birds in the art of Teotihuacan have been a problem for long time because of the difficulty of identifying them. The same birds have been identified as quetzals, owls, parrots, and eagles (Berrin, 1988, pp. 163–164).

Figure 4 shows one of many "small bird" images from this complex of murals. The bird's features suggest an amalgam of species, but the rich green plumage certainly suggests the Resplendent Quetzal. Who were the Aztecs? The last in a millennial Mesoamerican cultural lineage.

Ethnobiology

The study of ethnobiology is an effort to document the knowledge of the natural environment that is shared within a community, together with the wisdom that informs it and the practical applications of that knowledge to support local livelihoods (Anderson, 2011). Each such body of local knowledge constitutes a "folk biology." Ethno-ornithology simplifies the task by restricting the focus to the local knowledge of one well-defined group of living things, birds (the vertebrate class Aves in the Linnaean classification scheme).

Indigenous and/or subsistence-reliant local communities with deep attachments to local environments have received the most attention in ethnobiological research. Such communities are likely to have vibrant traditional environmental and ecological knowledge of local ecosystems. Thus, in documenting the cultural knowledge of such communities, we may best appreciate the universality of the human science of natural history.

In the last decade, I have been involved in a collaborative comparative ethno-ornithological project to compare the contemporary local knowledge of the bird life in a wide sample of Indigenous Mexican communities (Alcántara-Salinas et al., 2015). I contributed to this project by summarizing my research in two of these communities, one a Mayan-speaking town in highland Chiapas (Hunn, 1977), the other a Zapotec-speaking town in southern Oaxaca (Hunn, 2008). My colleague, Graciela Alcántara-Salinas has documented ethno-ornithological knowledge in three communities, in Zapotec and Cuicatec pueblos in northern Oaxaca (2011) and in a Nahua-speaking town in Veracruz. The ethno-ornithology of a Yucatec Mayan village is described by Anderson and Medina Tzuc (2005), that of the Comcáac or Seri by Morales Vera (2006), and that of the Pima Alta by Rea (2007). These local Indigenous communities span the Mexican nation. Together they provide haven for nearly 70% of Mexico's 1,100 bird species. Communities were studied in detail; each named between 80 and 150 local bird species.

Out of curiosity, I decided to add an Aztec perspective for contrast to these studies of these relatively contemporary societies and their bird knowledge. The Aztec Empire cannot be directly compared to contemporary Mexican Indigenous communities. At the time of the Spanish conquest (1521) the Aztecs encompassed not only rural peasant communities, but also an urban elite. Modern-day Mexican Indigenous communities share a Mesoamerican inheritance, including a fascination with birds of all kinds. The murals decorating inner walls at Teotihuacán, dating to between 250 BC and 800 AD, show how deep are the roots of this Indigenous world view, as they highlight the brilliant plumage of Mexican birds, likely Resplendent Quetzals and perhaps Lovely Cotingas, as well as predators such as eagles and owls.

Due to the early efforts of Fray Bernadino de Sahagún, we may now appreciate the Aztec fascination with birds as it informed the ancient Indigenous Mesoamerican science of natural history. Sahagún's decades-long collaborative ethnographic enterprise at the colonial *Colegio de la Santa Cruz* ultimately produced an encyclopedia in 13 books, treating of Aztec history, society, economy, religion, and of particular interest here, natural history. These books were dictated in the Mexica language, classical Náhuatl. The texts were recorded by descendants of the Aztec nobility who had been trained at the *Colegio* to write their native language using a Latinate script. Sahagún identified a number of these Aztec "multilingual grammarians and scribes" by name, but the Indigenous sources of the bird accounts are not known. Sahagún then composed a contemporary Spanish translation, which closely follows the Náhuatl text, with occasional gaps but also added context.

Several hand-written versions of the Náhuatl texts were produced in Mexico by 1577, though just one survived, the *Florentine Codex*, discovered centuries later in Florence, Italy. My focus here is on Book 11 of this *Codex*, which is devoted to "Earthly Things," of which the "Second Chapter, … telleth of all the different kinds of birds, of whatever sort " (Dibble & Anderson, 1963, hereafter cited as FC). More than 130 types of birds are named and described in the chapter, sometimes in striking detail. This number of species recognized and named is within the range of ethno-ornithological inventories from contemporary Indigenous Mexican communities. Colored illustrations accompanied these descriptions and were printed with the English translations in the Dibble and Anderson volume. Though these illustrations are of limited use in determining the identities of the birds named, I include copies of a number of the images to flesh-out the texts. However, my primary focus involves an analysis of the names and a careful comparison of the original descriptions with contemporary ornithological accounts to support well-educated guesses as to the various bird species described by these Aztec observers.

Some critics argue that words bear no necessary connection to real "things" of the world, that words are simply products of social convention, and therefore to attempt to equate the vocabulary of an Indigenous language with the Latinate designations of contemporary academic science is to distort the Indigenous reality beyond recognition (cf. Ellen, 1982, p. 206ff, for a discussion of this issue). I beg to differ. While there is never a perfect correspondence of Indigenous names for "species" of living things, it is inescapable that Indigenous observers experience the reality of "species" much as do contemporary field ecologists studying the dynamic interactions of "species" within a local biotic community. A "species" from the perspective of both the field ecologist and the Indigenous observer is a local population of organisms that share a role in the local "more-than-human" community and which "breed true."

Folk biology is a naked-eye science, an applied science, if you will, in the service of "making a living," but also much more, as it represents an engagement with the encompassing animate world. Folk biology does without the analytic power of modern evolutionary theory and the technical mastery of laboratory technologies such as genetic and biochemical analysis. Such contrasts of method and perspective, however, apply with far less force to field studies of specific local environments, such as might be pursued by ecologists. Field ecologists often recognize their local guides as highly perceptive and careful observers of the facts on the ground, often as colleagues (Diamond, 1966; Johannes, 1981; Majnep & Bulmer, 1977). The local experts who provided this Indigenous Aztec account of the "earthly things" must impress us with the sophistication of their local environmental knowledge—considering here just their ornithological acumen—the product of an Indigenous cultural tradition that owes nothing to colonial tutelage and that is in every respect on par with such European natural historians as Aristotle (LeRoi, 2014) and Linnaeus (Atran, 1990).

Tentative modern scientific translations of the Aztec bird names were offered by Rafael Martín del Campo of the Instituto de Biología at the Universidad Autónoma de México (1940). He suggested possible identities for 113 species, working from the Spanish edition, the *Historia General de las Cosas de Nueva España* (García Quintana & López Austin, 1988). His identifications have been generally accepted and were incorporated into the English translations of the *Codex* by Dibble and Anderson.

However, many of Martín del Campo's tentative Latin equivalents are clearly wide-of-the mark. I agree with his identifications for 78 of the 113 species for which he offered an opinion. However, I will suggest alternative identifications for 34 of the species he identified and offer "educated guesses" as to the identities of 19 additional birds (see Appendix 6). Our differences of opinion are not surprising, given that illustrated field guides were nonexistent before the 1970s and a great deal of new information has come to light with regard to Mexican bird distributions in the intervening 80 years. Between us we suggest identifications for 133 of the 135 birds named in the *Codex*. In any case, we can appreciate the sophistication of these Aztec observers of nature, considering the quality of detail provided for many of the birds named. Also, we can appreciate the diverse roles birds played in the Aztec *cosmovisión*.

The birds noted are at home in a wide range of habitats from throughout the empire. Given the imperial reach of the Aztecs in 1519, this is not surprising. The Aztec emperors commanded tribute from far-flung provinces, and brilliant feathers were among the most valued of tribute items. For example, the Resplendent Quetzal (QUETZAL-TŌTŌ-TL, *Pharomachrus mocinno*) nests in Tecolotlán province, literally, "land of owls," which is likely in what is now the Pacific Coast of Chiapas. The scribes described and named nine distinct types of plumes from this one special bird, ultimately to be woven in the sacred feather capes worn by the Aztec nobility, products of expert feather-workers of the *amanteca* guilds. The emperor exhibited his near-divine power, donning a stunning headpiece dominated by some 400 quetzal tail plumes (Figure 6).

The *Codex* devotes a concluding tenth paragraph "which telleth of the parts of the different birds" (FC, p. 54) documenting the expert knowledge of the master feather weavers.

Figure 6. Moteuczoma's headdress (with permission of the Kunsthistorisches Museum, Wien).

Sixteen feather types are named; wing feathers are distinguished as lesser, middle, and greater coverts, flight feathers and primaries, each feather tract separately named. The various "soft parts" are analyzed, noting how they vary across the diversity of local birds: the bill, the eye, the eyelid (that is, the nictitating membrane), the head, neck, tongue, throat, gullet, body form, joints, belly, intestines, crop, gizzard, even the rump gland, and the feet and toes. These details reveal an obsessive fascination with birds of all types, closely observed.

Of the 130-odd species named—given the challenge of establishing correspondences between Náhuatl bird names and modern Latin equivalents, we cannot confidently provide a more precise number—some 50 species were noted for their cultural values: 30 were deemed edible, 26 were valued for their feathers, 18 were judged "good singers" or were otherwise noted for their voices, seven marked seasonal changes, and another seven were of outstanding aesthetic value. Six were considered omens of ill-fortune, four had mythic value, four were domesticated, three had medicinal applications, while the Common Raven (*Corvus corax*) was noted as a crop pest. In a number of cases, ecological associations, migratory patterns, and other distinctive behaviors were described.

The great majority of the 130-some species named in the *Florentine Codex* are found in Central Mexico, in and around the Valley of Mexico and in the wider region known to the

Aztecs as Anáhuac. (The province of Anáhuac apparently extended to the coast of Chiapas.) The *Codex* details some 40 species of "water birds" (FC, pp. 26–39) residents on or visitors to the lakes surrounding the Aztecs' island capital, Tenochtitlán. This indicates the particular salience of the wetland habitats that dominated the Valley of Mexico in Aztec times. Many of these "waterfowl" were considered "edible, savory." They were hunted by "Water Folk," those who lived from harvests of the lakes that surrounded the capital city.

Eleven varieties of hummingbirds (HUĪTZIL-IN, Trochilidae) were named. These likewise contributed their brilliant jade and turquoise feathers for the *amanteca* artisans. It is notable that the Aztec patron god, a warrior deity, was known as *Huitzilipochtli* "Hummingbird-of-the-Left." The Aztecs, that is, the *Azteca*, were named for their legendary homeland, *Aztlan*, perhaps, "Land of the Snowy Egret" (AZTA-TL). The Feathered Serpent, Quetzalcoatl (Figure 7), is featured in Mesoamerican iconography from Teotihuacán to Chichen Itzá. Clearly, Aztec culture—and Mesoamerican culture more generally—was deeply appreciative of the beauty and power of birds. Twenty-six "birds of prey" are named, including eight kinds of "eagles" (CUĀUH-TLI, inclusive of the larger hawks) and eight of "falcons" (TLOH-TLI), though this total appears to include some synonyms.

Figure 7. Quetzalcoatl. Reproduced with permission of the Akademische Druck- u. Verlagsanstalt from Folio 22 in the *Codex Borbonicus*.

Contemporary academic ornithology seeks to classify every known species of bird within a multi-level taxonomic hierarchy within the vertebrate class, Aves, inclusive of a cascade of subordinate ranks: orders, families, genera, species, and a plethora of intercalated super-, sub-, and infra- ranks, all designed to best represent our present understanding of the evolutionary lines of descent of the world's birds. Local "folk ornithologists" are concerned, rather, with appreciating each salient local species in terms of its place in the local "economy of nature." Rather than attempt to place some 10,000 living bird species within a single hierarchic frame, they attend to the significance of 100 to 200 species of the several hundred that share their local life-space. The Aztecs, like other "folk scientists," deployed a shallow quasi-taxonomic framework. "Birds" constitute a "life-form" (Berlin, 1992), set apart as TŌTŌ-TL (plural, TŌTŌ-MEH), inclusive of all and only the birds named in this chapter. They were defined by their possession of feathers (IHUI-TL). There were also a few named "intermediate" groupings that help organize the Aztec bird list, as we will see below.

The Bird Chapter of the *Codex* is organized into 10 paragraphs (FC, pp. 19–56). These often seem rather arbitrary associations. I suspect this organization might have been encouraged by Sahagún, the colonial editor. We should also note that Sahagún employed questionnaires to guide the ethnographic interviews, which helps explain the often-programmatic quality of the descriptions. The first paragraph is not specifically labeled but seems to focus on birds of special value for their "precious plumage," such as the Resplendent Quetzal (QUETZAL-TŌTŌ-TL). The second paragraph details the parrots and hummingbirds, which, like the birds of the first paragraph, are notably colorful. The third paragraph, as we will see below, includes the "birds that live on the water" ("*tōtō-me, atlan nemi*," literally "birds, that live in water"). The fourth paragraph is devoted to "the birds of prey" ("*tlahuitequini*"). The fifth paragraph "telleth of still other kinds of birds, of whatever sort." The sixth paragraph "telleth of still other kinds of birds," but treats only quail. The seventh paragraph "telleth of still other birds, of their habits." The eighth paragraph "telleth of the birds which are good singers" ("*tōtō-me, huel cuica-ni*"), while the ninth is devoted to an extended account of the domestic turkey, and the tenth to an elaborate avian anatomy. The logic for most of these paragraphs is neither explicit nor obvious, indicative of the shallow taxonomic hierarchy employed by the Aztec scholars.

The Spanish version published as *Historia General de las Cosas de Nueva España* rather loosely follows the original Náhuatl, but at times offers useful elaborations missing in the *Codex*. For example, the first paragraph of "Capítulo II, *de las aves*," specifies that the first paragraph is "de las aves de pluma rica," a detail implicit in the original. The Spanish text elaborates on hunting the bird called XIUH-TŌTŌ-TL, literally, "turquoise bird," identified as the Lovely Cotinga:

> Esta ave se caza en el mes de octubre, cuando están maduras las ciruelas.
> Entonces las matan con cebretanas en los árboles. Y cuando caye a tierra,

arrancan alguna yerba para que tomándola no llegue la mano a las plumas, porque si llega dicen pierde la color (García Quintana & López Austin, 1988, p. 692). [This bird is hunted in the month of October, when the wild cherries are ripe. Then they kill them with blowguns in the trees. And when one falls to the ground, they pull up some herb so that picking it up their hand does not touch the feathers, because if they are touched they say they lose their color <author's translation>].

Rafael Martín del Campo's identifications were based on this Spanish version (1940).

The Method in My Madness

A careful reading of the *Florentine Codex* bird chapter in English translation leaves no doubt that the Aztec scribes and their expert sources were meticulous observers of the birds in their orbit. Considering the fact that these Indigenous ornithologists had no binoculars, no field guides, no cameras, no audio recording devices, and no alphabetic system of writing prior to learning the Latinate script, and that they lived their lives largely within the confines of the Valley of Mexico, what they accomplished in documenting the Mexican avifauna circa 1560 is, in my estimation, amazing. Imagine being asked to write detailed accounts of every kind of bird you had ever seen from memory. Yet they produced a thesis of some 6,000 words in their native Náhuatl language naming and describing some 130 kinds of birds. Nevertheless, the *Codex* descriptions can be and often are misleading, either by omission, by commission, or simply by editorial errors. Illustrative examples might include the description of TZINITZ-CAN, by consensus the Mountain Trogon, as a "waterfowl," which it is not by any stretch of the imagination; or the statement that the bill of XOCHI-TENACAL, the Keel-billed Toucan, is "concave" when it is obviously "convex," likely a translation error as the Náhuatl term in question, TEN-ACAL-TIC, is better translated as "boat-shaped" (Karttunen, 1983, p. 1).

To better appreciate the quality of this Indigenous science we need a detailed comparison of the Aztec knowledge of birds with that of contemporary observers. To facilitate such a comparison, we are fortunate to have available a comprehensive *Guide to the Birds of Mexico and Northern Central America* (Howell & Webb, 1995), fully illustrated—when complemented by North American field guides (e.g., *The Sibley Guide to Birds, Second Edition* [Sibley, 2014])—and with detailed range maps for each species known to have occurred in Mexico and northern Central America (some 1,070 species), showing seasonal distributions, with notes on habits, vocalizations, habitat, and elevation.

As noted above, complementing the *Codex* is a 16th-century Spanish translation of the original Náhuatl text, the *Historia General de las Cosas de Nueva España* (García Quintana & López Austin, 1988). The Mexican ornithologist, Rafael Martín del Campo (1940), based his identifications on the Spanish texts of the *Historia General*. His suggestions were by-and-large accepted by the authors of the first definitive "Distributional Check-list of the Birds of Mexico" (Friedmann et al., 1950; Miller et al., 1957) and by Dibble and Anderson for their English translations of bird names. However, we have learned a great deal about Mexican bird life since then (cf. Howell & Webb, 1995). With this new information, I believe we are in a much better position today to interpret these classic texts. In addition, the academic study of the Náhuatl language of the Aztecs has advanced and rich lexicographic resources are now readily available that facilitate an analysis of these Aztec bird names (Karttunen, 1983; Wood, 2000–2020).

I also had a particular advantage when it comes to comparing the *Codex* to contemporary ecological texts. I have studied Indigenous systems of biological nomenclature and classification for the past 50 years, including detailed field studies with Mayan and Zapotecan languages in southern Mexico. I have been an avid birder for nearly 60 years and have a personal acquaintance with the great majority of Mexican birds. It is well known that to fully appreciate an aspect of traditional environmental knowledge, the researcher benefits greatly from sharing with his Indigenous partners a passion for the subject matter. I rely on this experience to inform my reading of these classic Náhuatl natural history texts.

The trick is to match as best we can the Náhuatl descriptions of some 450 years ago with our current knowledge. We must assume, of course, that the birds in question have not drastically altered their behaviors or distributions in the interim. In a few cases we know of significant changes, such as the extinction of the Imperial Woodpecker (*Campephilus imperialis*) of the highland pine-oak forests of northwestern Mexico (last reported 1956) and of the Slender-billed Grackle (*Quiscalus palustris*), last reported 1910, extirpated from its limited original range in the wetlands at the headwaters of the Lerma River, since largely drained (Howell & Webb, 1995). A few species are known to have retreated south from their historic ranges, e.g., the Scarlet Macaw (*Ara macao*), the King Vulture (*Sarcoramphus papa*), the Harpy Eagle (*Harpia harpya*), and the Red-throated Caracara (*Daptrius americanus*). Other species are recent introductions, such as the Cattle Egret (*Bubulcus ibis*), Rock Pigeon (*Columba livia*), and House Sparrow (*Passer domesticus*).

Curiously, the *Florentine Codex* described how at the behest of the Emperor Ahuitzotl (1486–1502), the Great-tailed Grackle (*Quiscalus mexicanus*) was introduced into the Valley of Mexico from its original homeland in the coastal lowlands of Veracruz, because "It has a good voice; it speaks well, it speaks pleasantly" (FC, p. 50). (Not to my ear!) It is possible that the now-extinct Slender-billed Grackle could have resulted from a much earlier human intervention, as its highland habitat is as anomalous as that of the Great-tailed. The Aztecs also raised a variety of species of parrots and parakeets, perhaps altering their natural distributions in the process, though there are no positive records of such range extensions. The disjunct population of the Military Macaw in northeastern Mexico might be explained as due to the well-documented Aztec trade in exotic birds (Haemig, 1978). All things considered, I believe it is safe to assume that contemporary avian distributions closely approximate those of Aztec times, unless proven otherwise.

I also assume that the "birds" named are actual birds rather than mythologized creatures, such as the Phoenix of Mediterranean mythology or the Thunderbird of Pacific Northwest Indian traditions. This is a working hypothesis, not axiomatic. Likewise, I assume that each bird named is a unique entity, so that, once named it is not likely to crop up again unannounced under a different name. In a few cases, two quite different birds, judging by their descriptions, were identified as the same species by Martín del Campo. Ā-CIH-TLI and ĀCA-CHICHIC-TLI were both identified as the Western Grebe. While the former fit the descriptive account, the latter did not. Rather, the Least Bittern is a much closer match for ĀCA-CHICHIC-TLI.

Of course, a bird may be known by two or more names, cases of synonymy. But the Aztec scribes were quite careful to note such instances. It is also to be expected that males, females, and perhaps young of certain species of particular value may be given gender- and/or age-specific names, as in English we have "rooster" and "hen." This is the case in the *Codex* for the Yellow-headed Parrot (young are known as TOZ-NENE, adults as TOZ-TLI), quail (ZŌL-IN for the species, TECU-ZŌLI for the cock, and OHUA-TON for the hen) and the House Finch (MOLO-TL for the species, CUA-CHICHIL or NŌCH-TŌTŌ-TL for the red-headed male). In several instances a single name may have two (or in one case, three) distinct but related senses. This is known as polysemy. A prime example is CANAUH-TLI, which based on the varied descriptions under sequential headings in the *Codex* may refer to ducks in general (CANAUH-TLI$_1$), or to one or two "prototypical ducks." These are most likely the Mexican Duck (*Anas diazi*, CANAUH-TLI$_2$), on the one hand, and the Muscovy Duck (*Cairina moschata*, CANAUH-TLI$_3$), on the other. Likewise, CUĀUH-TLI$_1$ names eagles and large hawks in general or, in context, specifies the Golden Eagle (*Aquila chrysaetos*, CUĀUH-TLI$_2$), which may also be named ITZ-CUĀUH-TLI, literally, "obsidian eagle."

The descriptions in the *Codex* vary from a few brief words to lengthy essays noting the type of bird, its habitat and range, extensive plumage and soft-part details, behaviors, vocalizations, and seasonal movements. The more elaborate descriptions may also describe how local people understood the significance of a bird for their lives, either for food, medicine, materials, aesthetic appreciation, and ecological indications, as well as their powers as omens, auguries, and dangers. The name or names assigned each bird may also reveal key details that may help in the identification process. The names, thus, may have "descriptive force" (Hunn & Brown, 2011).

Binomial compound names often highlight key contrasts among similar species. For example, METZ-CANAUH-TLI, literally, "moon duck," is the Blue-winged Teal, the male of which sports a unique white crescent-moon shaped face-patch. Its close relative, the Cinnamon Teal, is named CHIL-CANAUH-TLI, literally, "red duck." Such clues can only hint at the bird's identity in most cases. CUAPPACH-HUĪTZIL-IN, literally, "tawny hummingbird," is likely the Cinnamon Hummingbird (*Amazilia rutila*), while CHALCHI-HUĪTZIL-IN, literally, "turquoise hummingbird," may be the Mexican Violetear (*Colibri mexicanus*), contrasting in terms of their dominant plumage color. However, given the diversity of Mexican hummingbirds, these identifications remain provisional, as other identifications are possible. The prototypical owl, TECOLŌ-TL, is the Great Horned Owl, its deep hoots nicely mimicked in the name. ZACA-TECOLŌ-TL, literally, "grass owl," is likely the Short-eared or Striped Owls (*Asio flammeus* or *A. clamator*), which may be common in their prairie habitat.

Attempts at translating the Náhuatl bird names into modern scientific Latin and English often involve "educated guessing." Range, habitat, size and characteristics of plumage, beak, and legs, and very often vocalizations allow elimination of many potential species from consideration, perhaps leaving just one well-justified candidate. In such cases, we may conclude that, for example, TOL-COMOC-TLI is the American Bittern, a resident breeding in the Valley of Mexico reedbeds that sounds like a two-toned log drum. It could be nothing else. A significant fraction of the birds of the *Codex* are identified with comparable certainty.

On other occasions we may eliminate all but two or three species, any one of which might fit the details, but one of which is somewhat more likely on distributional grounds. This is the case for TZINITZCAN-TŌTŌ-TL, which Martín del Campo and I agree most likely referred to the Mountain Trogon. Two other similar species were also likely known to the Aztecs, the Elegant and the Collared Trogons. However, the Mountain Trogon is the characteristic species of the highland pine-oak forests that surround the Valley of Mexico and thus most likely to have been seen regularly. In this case it is reasonable to assume that the name could be applied to any or all three of these trogons, a case of an "extended range" of a category.

Finally, we may be faced with a considerable number of possibilities, as in the case of a hummingbird that is "resplendent," which might describe dozens of hummingbird species, or for which the details recorded are so sketchy or confounding that we can only speculate as to its identity. TEN-ITZ-TLI is a case in point. It is said to be a nocturnal bird the size of a dove that hawks flying ants and that has three beaks and two tongues. I know of no bird that fits that description but, making allowance for the likelihood that such a bird was rarely observed, it could have been a Black Tern, a not totally unreasonable possibility, though it does not have three beaks (nor does any bird). I qualify such identifications as "possibly" or "perhaps" the bird named.

The Birds of the *Florentine Codex* Deciphered

I analyze each bird named in the *Codex* more-or-less in the order of the original. Each entry opens with an all-caps canonical version of the name, modified to reflect the best available orthographic representation, including marking vowels as long or short. These canonical forms match as far as possible the forms listed alphabetically by Karttunen in his *Analytic Dictionary of Nahuatl* (1983) and as represented in the *Online Nahuatl Dictionary* (Wood, 2000–2022). If the name has a transparent reading, I then suggest its literal sense, or, if it is clearly onomatopoetic, that is, if the name mimics a characteristic vocalization, that will be noted. Following that, I indicate the identity of the bird or birds named, first with their currently accepted English name, prefaced by "perhaps" or "possibly" if the identification is uncertain. The Latin name follows in parentheses. Next, I quote selections from the English translation of the Náhuatl text from the *Florentine Codex* (Dibble & Anderson, 1963).[1] I indicate the *Florentine Codex* page source as (FC, p. xy). I enclose my editorial clarifications in angle brackets to distinguish those clarifications from those of the *Codex* editors in square brackets. Finally, I review the evidence for one or another identification and may offer additional commentary.

"First Paragraph," Birds with Rich Plumage

We begin with the "First Paragraph, which telleth of the many different kinds of birds, of whatever sort" (FC, p. 19). However, the 13 birds named here appear to share a special place by virtue of their highly valued and notably colorful plumage. This is explicit in the Spanish version, the *Historia General de las Cosas de Nueva España* (García Quintana & López Austin, 1988): "*Párrafo primero*, de las aves de pluma rica" <"First paragraph, of the birds of rich plumage"> (Garcia Quintana & Lopez Austin, 1988, p. 690). Most of these did not occur in the Valley of Mexico, but both live birds and their feathers were obtained either in trade by the professional merchant class, the *pochteca*, or as tribute from satellite states to be woven by the *amanteca* feather-workers into the rich capes and headdresses privileged by the elite.

---

1 The School of American Research granted permission to use the English translations of the original Náhuatl text. These translations are quoted from the Fray Bernardino de Sahagún in C.E. Dibble & A.J.O. Anderson 1963. Copyright 1981. All rights reserved.

Eugene S. Hunn

QUETZAL-TŌTŌ-TL, literally, "Quetzal-bird," Resplendent Quetzal (*Pharomachrus mocinno*) (Figures 8 and 9):

> Its bill is pointed, yellow; its legs yellow. It has a crest, wings, a tail. It is of medium size, the same as the slender-billed grackle…. On [the tail], the feathers which grow on it are called *quetzalli*. Those which are on its tail are green, herb-green, very green, fresh green, turquoise-colored. They are like wide reeds: the ones which glisten, which bend…. This bird is crested; of quetzal spines, of quetzal thread feathers is its crest, very resplendent, very glistening. They are called *tzinitzcan*…. About its neck, its throat, and its breast, [the feathers] are reddish…. Chili-red, resplendent, wonderful, precious. The name of the feathers is *tzinitzcan*…. The breeding place of these birds is [the province of] Tecolotlan <literally, "Land of Owls">; and in the trees they make their homes and raise their young (FC, pp. 19–20).

This is the first bird named in the *Codex*. It is unmistakable. Its brilliant emerald plumes were offered in tribute by Aztec dependencies south of the Isthmus of Tehuantepec in what is now the Mexican state of Chiapas and neighboring Guatemala, that is, "the province of Tecolotlan." "*Quetzal*" names the rich plumes rather than the bird. The royal headdress of the

Figures 8 and 9. QUETZAL-TŌTŌ-TL, Resplendent Quetzal (ML433742111/121, photos by Heber David Díaz Gutiérrez from Macauley Library).

emperor Moteuczoma Xocoyotzin—who met the conquistador Cortés in the Aztec capital in 1519—sported a broad fan of some 400 quetzal tail plumes ("Moteuczoma's headdress," held at the Kunsthistorisches Museum, Wien). These precious feathers are also called *tzinitzcan*, namesake of our next subject.

TZINITZCAN-TŌTŌ-TL, literally "*Tzinitzcan* bird"/TEŌ-TZINITZCAN, literally, "sacred *tzinitzcan* plumage," the Mountain Trogon (*Trogon mexicanus*) (Figure 10): "Its feathers are black, dark. And for this reason it is called *teotzinitzcan*: on its breast and its underwing it is varicolored, half black, half green. It is glistening green, resplendent" (FC, p. 20). As noted for the Quetzal, *tzinitzcan* feathers may be either green or red, but in any case, are "resplendent." Martín del Campo identified this bird as one or another of the three central Mexican species of red-bellied trogons (each of which replicate the palette of the quetzal), but most likely focused on the Mountain Trogon. The other possibilities are the Elegant Trogon (*Trogon elegans*) and the Collared Trogon (*Trogon collaris*). They are less partial than the Mountain Trogon to the high elevation pine-oak forests surrounding the Valley of Mexico but were likely included in the extended range of this category. There is one puzzling element of the *Codex* description of TZINITZCAN-TŌTŌ-TL. It opens by stating that, "It lives in the water." I suspect this is an error introduced in the original, as no trogon is particularly associated with water and certainly not the Mountain Trogon.

Figure 10. TZINITZCAN-TŌTŌ-TL, Mountain Trogon (ML461204061, photo by Gabriel Cordón, from Macauley Library).

TLĀUH-QUECHŌL/TEŌ-QUECHŌL, Roseate Spoonbill (*Platalea ajaja*) (Figure 11): "Also its name is *teoquechol*. It is a waterfowl, like the duck: wide-footed, chili-red footed…. It is wide-billed; its bill is like a palette knife…. [Its] plumage becomes pale, pink, chili-red, well textured" (FC, p. 20). It is included here rather than later, in the third paragraph which is devoted to the waterfowl, no doubt due to the particular value of its colorful pink plumage. The modifier TLĀUH- indicates firelight and red ochre (Karttunen, 1983, p. 270); TEŌ- means "sacred." Karttunen (1983, p. 206) notes that, "QUECHŌL refers not to the color of the bird but apparently to the characteristic sweeping motion of its neck <when feeding>." There are just two possibilities, the Roseate Spoonbill and the American Flamingo (*Phoenicopterus ruber*). The spoonbill is the best fit based on its distribution along both Mexican coasts, while the flamingo is restricted to the Caribbean littoral of the Yucatan peninsula. The colorful spoonbill feathers were—like those of the quetzal—prized tribute from coastal dependencies.

Figure 11. TLĀUH-QUECHŌL, the Roseate Spoonbill (Dibble & Anderson, 1963, Figure 44, after Paso y Troncoso).

Two other birds named in this "First Paragraph" share the head term QUECHŌL, XIUH-QUECHŌL and XIUH-PAL-QUECHŌL. Both are most likely motmots (Momotidae). Martín del Campo identified the first as Lesson's Motmot (*Momotus lessonii*) and the second as the Turquoise-browed Motmot (*Eumomota superciliosa*). As noted above, Karttunen stated that, "QUECHŌL refers not to the color of the bird but apparently to the characteristic sweeping motion of its neck <when feeding>." This is likely relevant also to the motmots, as they characteristically "twitch their tails from side to side" (Howell & Webb, 1995, p. 437). The *Codex* descriptions fail to mention this, which to me is perhaps the most distinctive feature of the motmots, their "racquet tails," which they switch back and forth like the pendulum of a clock. I concur with Martín del Campo's identifications given their colorful plumage, which no doubt brought them to the attention of the Aztec observers, obsessed as they were with the beauty of "resplendent" feathers, notably shades of brilliant blue and green.

XIUH-QUECHŌL > XIHU(I)-TL, "grass or green stone, turquoise," Lesson's Motmot (*Momotus lessonii*): The accompanying details in the *Codex* are sparse: "Its feathers are herb green; its wings and tail are blue. It lives in Anahuac" (FC, p. 20). The Spanish editors of the *Historia General* specified that Anahuac lies east of [the Valley of] México toward the Southern Sea (García Quintana & López Austin, 1988, p. 691). Lesson's Motmot is common on the Atlantic slope forests of Central Mexico.

XIHUA-PAL-QUECHŌL possibly > XIUH-PAL- "for, by means of turquoise"/TZIUH-TLI, Turquoise-browed Motmot (*Eumomota superciliosa*) (Figure 12): "Its name is also *tziuhtli*. The bill is long. The legs are black. Its head, and its back, and its wings, and its tail are light blue; its belly and its wing-bend tawny." The Turquoise-browed Motmot is restricted to the Yucatan Peninsula and the coast of Chiapas. It is the only species with a "tawny belly," and despite its more limited distribution, is common at the eastern limits of "Anáhuac," as defined in the *Historia General*.

Figure 12. XIHUA-PAL-QUECHŌL, Turquoise-browed Motmot (ML434528501, photo by Guillermo Saborío Vega from Macaulay Library).

Three species in this paragraph share the element -CUAN. ZA-CUAN, the Montezuma Oropendola (*Psarocolius montezuma*), and two quite different birds, both named AYO-CUAN. AYO-CUAN$_1$ is apparently the Yellow-winged Cacique (*Cacicus melanicterus*), a relative of the oropendola just mentioned, while AYO-CUAN$_2$ is a water bird. I believe this is the Northern Jacana (*Jacana spinosa*). I have been unable to analyze these names, though the fact that they are listed in order in the *Codex* suggests some common semantic relationship.

ZA-CUAN, the Montezuma Oropendola (*Psarocolius montezuma*) (Figure 13):

> It is pointed of bill;… Everywhere [over the body] its feathers are tawny. And for this reason it is called *çaquan*; its tail is yellow, very yellow, intense yellow,… but there are black [feathers] which cover it. When it spreads its tail, then the yellow shows through (FC, pp. 20–21).

The account includes a puzzling note, that "the feathers over its nose are chili-red" (FC, p. 20). No such feathering is evident in the plates in Howell and Webb (1995, Plate 65.10). Nevertheless, there is no doubt about this identification. A second species of oropendola, the Chestnut-headed (*Psarocolius wagleri*) might be included within the extended range of ZA-CUAN, as it is quite similar to the Montezuma Oropendola, though its range is more southerly.

Figure 13. ZA-CUAN, Montezuma Oropendola (ML435008871, photo by Carolien Hoek from Macaulay Library).

AYO-CUAN$_1$, Yellow-winged Cacique (*Cacicus melanicterus*) (Figure 14):

> It is a forest dweller... [in the province of] Cuextlan and in Michoacan
> <west of the Valley of Mexico toward the Pacific Ocean>.... The bill is
> pointed, black; everywhere [over the body] its feathers are black, but its tail
> is mixed white <*sic.*> [and black], so that it is called *ayoquan* (FC, p. 21).

This suggests that ZA-CUAN and AYO-CUAN refer, respectively to the predominant color of the body plumage of these two species, "tawny" and "black," respectively. We should note that the Yellow-winged Cacique's tail is mixed yellow and black, not white and black as stated in the *Codex*. Nevertheless, as Martín del Campo argued, there is no other reasonable possibility, given the other details.

Figure 14. AYO-CUAN$_1$, Yellow-winged Cacique (ML432006541, photo by Suzanne Roberts from Macaulay Library).

AYO-CUAN$_2$, possibly the Northern Jacana (*Jacana spinosa*) (Figure 15):

> It is a water bird. Thus does it live: accompanying it go all the water birds ….
> It is yellow-billed, green of wing-bend; its flight feathers, its tail are [as if]
> shot with mirror-stones …. Everywhere [over its body] its feathers are rud-
> dy (FC, p. 21).

Martín del Campo suggested this might be the Agami Heron (*Agamia agami*). However, that heron is rather rare and solitary and is restricted to the Atlantic lowlands, nor does it have a yellow bill. Though AYO-CUAN$_2$ differs radically from AYO-CUAN$_1$ in form and habitat, it bears a striking, if superficial resemblance, to both the oropendola and the cacique. Its body is black, contrasting with a "tawny" back, yellow bill, and chartreuse flight feathers. The *Codex* notes these as "green of wing-bend" (FC, p. 21). All three are striking birds that one would expect to stand out "from the crowd," for the Aztecs as they do for us.

Figure 15. AYO-CUAN$_2$, Northern Jacana (ML435268441, photo by Michael Car-
mody from Macaulay Library).

Next mentioned are four species sharing the head term TŌTŌ-TL "bird." The honey-creeper and the grosbeak are small forest birds, brightly colored. Their descriptions are brief, so one cannot be certain of the identification, given that colorful forest birds are not uncommon in Central Mexico, though I see no reason to dispute Martín del Campo's identifications. The cotinga and the Squirrel Cuckoo are quite distinctive and thus readily identified. Both contributed feathers for the emperor Moteuczoma's regal headdress (Figure 6 above, from the Kunsthistorisches Museum, Wien).

CHALCHIUH-TŌTŌ-TL, literally, "jade bird," likely the Red-legged Honeycreeper (*Cyanerpes cyanea*): "It is a forest dweller; small, pointed and small of bill. Its head and tail are herb-green,… And the under part of its wings, and all its body are light blue,… the color of fine turquoise" (FC, p. 21). This description may conflate the blue male with the green female of this species (cf. Martín del Campo, 1940, p. 389).

XIUH-TŌTŌ-TL, literally, "turquoise bird," by consensus, the Lovely Cotinga (*Cotinga amabilis*) (Figure 16): "It is an inhabitant of Anahuac,… Its breast is purple, its back a really light blue" (FC, p. 21). The Lovely Cotinga shows the unique light blue and deep purple color scheme described. The Spanish account in the *Historia General* elaborates:

> Esta ave se caza en el mes de Octubre, cuando estan maduras las ciruelas. Entonces las matan con cebretanas en los árboles. Y cuando caye a tierra, arrancan alguna yerba para que tomándola no llegue la mano a las plumas, porque si llega dicen pierde la color. <"This bird is hunted in the month of October, when the wild cherries are mature. Then they kill them with blowguns in the trees. And when they fall to the ground, they <the hunters> pull up a plant so that picking it up their hand does not touch the feathers, because they say if <they touch the feather> it loses its color"; author's translation> (García Quintana & López Austin, 1988, p. 692).

Figure 16. XIUH-TŌTŌ-TL, Lovely Cotinga (ML435428891, photo by Guillermo Saborío Vega from Macaulay Library).

ĒLŌ-TŌTŌ-TL, literally, "green maize bird," possibly the Blue Grosbeak (*Passerina caerulea*): "Its wings and its bill are of dull colors. It is [the color of] the lovely cotinga – light blue. It looks dull; it turns dull" (FC, p. 22). This is one of a number of small, predominantly blue birds. However, it is a common resident across the highlands of Central Mexico while the other species tend to have more restricted ranges or to be migrants. This grosbeak also has dark wings with brown wing-bars and a gray bill, fitting the description.

CUAPPACH-TŌTŌ-TL, literally, "tawny bird," the Squirrel Cuckoo (*Piaya cayana*) (Figure 17): "It is tawny: smoky, even-colored, well textured…. it is smoky" (FC, p. 22). Its plumes have been identified as a decorative element on the emperor Moteuczoma's regal headdress (Figure 6 above, from the Kunsthistorisches Museum, Wien).

Figure 17. CUAPPACH-TŌTŌ-TL, Squirrel Cuckoo (ML431536351, photo by Julien Amsellem from Macaulay Library).

Rounding out the list in the First Paragraph is the toucan:

XOCHI-TENACAL, literally, "flower beak," prototypically the Keel-billed Toucan (*Ramphastos sulphuratus*) (Figure 18), though perhaps extended to include the Emerald Toucanet (*Aulacorhynchus prasinus*), as Martín del Campo suggested. One detail offered in the Náhuatl account would seem to clinch my identification: "there are marks [on its bill] as of hawk scratches" (FC, p 22). That is a distinctive detail of the Keel-billed Toucan (Howell & Webb, 1995, Plate 34.1). The description of the bill shape as "concave" is a mistranslation, as it is rather obviously convex. The Náhuatl term in question is "*tenacaltic*" (TEN-ĀCAL-TIC), which is more accurately rendered "boat-shaped bill," cf. ĀCAL-LI "boat" (Karttunen, 1983, p. 1). A second problematical note in the *Codex* description is the assertion that, "it raises its young in the trees. It merely builds a bag-like nest for its young [and] suspends them" (FC, p. 22). It would seem the Aztec scribes confuse here the nests of oropendolas and caciques with that of the toucan, as the toucan, by contrast, nests "in unlined tree cavities" (Howell & Webb, 1995, p. 447).

Figure 18. XOCHI-TENACAL, Keel-billed Toucan (ML435008281, photo by Carolien Hoek from Macaulay Library).

"Second Paragraph," Parrots, Hummingbirds, and Others

The Second Paragraph lists five species of the parrot family (Psittacidae), 11 varieties of hummingbirds (Trochilidae), a wood-rail (Rallidae), two tinamous (Tinamidae), plus a bird of uncertain identity that "the people there <"in the province of Teotlixco, toward the southern sea"> say thus: that when we die, our hearts turn into [these birds]" (FC, p. 25). The challenge here is to determine how the Náhuatl names map to a wealth of local species, including 19 Psittacids, 55 species of resident hummingbirds (plus six winter visitors), and four tinamous. Not all are equally likely, so it is possible to offer educated guesses in a number of cases, but not all, as we will see.

TOZ-TLI, pride of place among the parrots goes to the Yellow-headed Amazon (*Amazona oratrix*):

> It has a yellow, curved bill, like that of the [COCHO]; the head is crested. Its breeding place is especially [the province of] Cuextlan.... When the young yellow-headed <amazon> [TOZ-NENE] is already developed, it turns yellow,.... It develops fluffy feathers. When completely feathered, then it is called *toztli* <TOZ-TLI, cf. TOZQUI-TL "throat, voice" (Karttunen, 1983, p. 249)> (FC, pp. 22–23).

The young are captured to be raised as household pets, admired for their "teachability" and vocal range, as well, no doubt for their bright green and yellow plumage. These feathers are known specifically as *xollotl*.

Two other amazon parrots are named, of which five might have been familiar to the Aztecs:

COCHO, most likely the White-fronted Amazon (*A. albifrons*) (Figure 19):

> It resembles [TOZ-NENE]. It has a yellow-curved bill; it is crested. Everywhere [over its body] its feathers are dark green; its coverts are dark red [and] dark yellow. Its feathers named *xollotl*, the small feathers of its tail [and] its wing, are ruddy. It is a singer, a constant singer, a talker, a speaker, a mimic, an answerer, an imitator, a word-repeater. It repeats one's words, imitates one, sings,... chatters, talks (FC, p. 23).

This emphasis on the value of "speech" is characteristic of the Aztecs. Note that the emperor's title was TLAHTOĀNI, cf. TLAHTOĀ "to speak, to issue proclamations and commands, for birds to chatter" (Karttunen, 1983, p. 266).

Figure 19. COCHO, White-fronted Amazon (ML464384391, photo by Margaret Barrow-Smith from Macaulay Library).

TLALACUEZA-LI, likely the Red-crowned Amazon (*A. viridigenalis*) (Figure 20): "a forest dweller,… [with] a chili-red head, purple-brown wing-bend, dark yellow breast" (FC, pp. 23–24). The Red-crowned Parrot is the only Central Mexican amazon parrot with a prominent red crown, so this equation is most likely. Three additional *Amazonas* species might be included within the extended range of this category or of COCHO, the Lilac-crowned Amazon (*A. finschi*), the Red-lored Amazon (*A. autumnalis*), and the Mealy Amazon (*A. farinosa*).

Figure 20. TLALACUEZA-LI, Red-crowned Parrot (ML434035091, photo by Kris Janicki from Macaulay Library).

QUILI-TON/ QUILI-TZIN, likely onomatopoetic with a diminutive ending, most likely the Orange-fronted Parakeet (*Psitticara canicularis*): "It resembles <TOZNENE> and <CO-CHO>. It is small, tiny; the small head is chili-red. Everywhere [the body is] herb-green, dark green …. Its food is maize …. I give it grains of dried maize to eat" (FC, p. 23). Which of several Central Mexican parakeets is the most likely referent is debateable, as the description in the *Codex* is brief. There are two species of the genus *Psitticara*, one resident in the Pacific lowlands, the Orange-fronted Parakeet, the other on the Atlantic side, the Aztec Parakeet (*Eupsittula* [previously *Psitticara*] *astec*). Neither shows a "chili-red head," though the Orange-fronted does flash orange on the forecrown, so might be the most likely candidate. Other parakeets of different genera (*Aratinga, Forpus, Brotogeris, Bolborhynchus*) seem less likely, and none has a red crown, though any and all might be included in the extended range. One authority voted for the sparrow-sized Mexican Parrotlet (*Forpus cyanopygius*) of the Pacific lowlands. However, the description suggests rather a more robust bird.

ALO, Scarlet Macaw (*Ara macao*) (Figure 21):

> It lives especially in [the province of] Cuextlan <that is, coastal Veracruz>, in crags and in the dense forest. It is tamable …. Flaming red are its eyes; yellow are its breast and belly …. Its tail, its wing [feathers] are ruddy, reddish …. They [these feathers] are called *cueçalin*, literally, 'tongue of fire'. The wing-coverts and tail coverts are blue, becoming ruddy, reddish, bright reddish, orange (FC, p. 23).

Scarlet Macaws were highly valued throughout Mesoamerica dating to centuries before the Aztec ascendancy, as there was active trade in feathers and captive macaws as far north as what is now the southwestern United States (Minnis et al., 1993).

Figure 21. ALO, Scarlet Macaw (ML434733041, photo by Janet Stevens from Macaulay Library).

CUAUH ALO, literally, "forest macaw," Military Macaw (*Ara militaris*), "*quauh alo. papagayo grande y verde*," <"a large green parrot"> (Molina, 1571, f. 86r. col. 2): not mentioned in the *Codex* but noted in the 16th-century Náhuatl dictionary of Fray Alonso de Molina.

Hummingbirds were known variously as:

HUĪTZIL-IN/ HUĪTZITZIL-IN/HUITZACA-TZIN/HUĪTZTZĪTZI-QUI, all onomatopoetic, all imitative of the buzzing of these tiny birds in flight, with various diminutive suffixes appended to emphasize their small size. Hummingbirds are classified in their own family, the Trochilidae (Figure 22): "The bill is … slender, small and pointed, needle-like…. Its food is flower honey, flower nectar. It is whirring, active [in flight] …. It flies, darts, chirps" (FC, p. 24).

The Aztec scribes were attentive to the diversity of species and plumages within this large bird family, naming 11 varieties. Ten are named binomially, with the generic head term HUĪTZIL-IN variously modified. Martín del Campo attempted to identify several of these hummingbird species, but his suggestions are clearly guesses, often wide of the mark, in my opinion. It is important to note that male and female plumages within a species often vary dramatically and that the young of the year may resemble the typically non-descript females. This is particularly true for species that breed to the north and visit Mexico in the winter, such as the Ruby-throated (*Archilochus colibri*), Black-chinned (*A. alexandri*), Rufous (*Selasphorus rufus*), Allen's (*S. sasin*), Calliope (*S. calliope*), and Costa's (*Calypte costae*) hummingbirds. For this reason, I am skeptical of Martín del Campo's identification of several of the Aztec hummingbirds as such migratory species as Costa's (XI-HUĪTZITZIL-IN), Allen's (TLE-HUĪTZIL-IN), and Ruby-throated (TOZCATLE-TON).

Figure 22. HUĪTZITZI-LI, hummingbirds (Dibble & Anderson, 1963, Figure 61, after Paso y Troncoso).

The Aztecs noted the seasonal scarcity of these birds in the highlands—dependent as they are on nectar and insects—attributing this to their "hibernation," hanging in trees from their beaks, shedding their feathers, to be revived with the onset of the rains. It seems they mistook the chrysalis of a butterfly or moth for a hibernating hummingbird. Peak hummingbird activity in the highlands is near the end of the rains in fall when flowers are most abundant.

Following are the "species" of hummingbirds listed in the *Codex*, typically with very brief descriptions:

QUETZAL-HUĬTZIL-IN, literally, "quetzal hummingbird," perhaps the Garnet-throated Hummingbird (*Lamprolaima rhami*): "Its throat is chili-red, its wing-bend ruddy. Its breast is green. Its wings and its tail [feathers] resemble quetzal feathers" (FC, p. 24). Martín del Campo voted here for the Broad-tailed Hummingbird, *Selasphorus platycercus*, which is a decent fit and a common resident in the Central Mexican highlands. However, the Garnet-throated Hummingbird, also a common highland resident, is a better fit, with its garnet gorget and bright rufous wings, which is my choice.

XI-HUĬTZITZIL-IN, literally, "turquoise hummingbird," perhaps the Broad-billed Hummingbird (*Cynanthus latirostris*): "It is entirely, completely light blue like a cotinga <that is, XIUH-TŌTŌ-TL>, pale like fine turquoise. It is resplendent like…fine turquoise" (FC, p. 24). Certainly not Costa's Hummingbird, Martín del Campo's suggestion, as Costa's is never "completely light blue," certainly not in winter when it visits northwest Mexico. The Broad-billed Hummingbird best fits the details described and is a common resident in the Central Highlands.

CHALCHI-HUĬTZIL-IN, literally, "jade hummingbird," possibly the Mexican Violetear (*Colibri mexicanus*): "Is light green; a turquoise shade; herb green" Martín del Campo voted here for the Broad-billed Hummingbird (see above), however there are a number of common Central Highland species that match this description. I lean toward the Mexican Violetear (*Colibri mexicanus*), a common and conspicuous highland resident.

YAUHTIC HUĬTZIL-IN, literally, "dark hummingbird," a species of hummingbird (Trochilidae): "The Yiauhtic uitzilin is dark green" (FC, p. 24). This might refer to any number of Central Mexican hummingbird species, so an educated guess is unwarranted.

TLAPAL-HUĬTZIL-IN, literally, "red dyed hummingbird," likely the Rufous Hummingbird (*Selasphorus rufus*): "is red and black" (FC, p. 24). Martín del Campo and I agree that this is quite likely the Rufous Hummingbird, though that species is a winter visitor and may lack the bright adult male coloration at that season.

AYOPAL-HUĪTZIL-IN, literally, "purple hummingbird," perhaps the Violet-crowned Hummingbird (*Amazilia violiceps*): "is light brown, the color of *tunas* <prickly-pear cactus fruits, which can be quite colorful>" (FC, p. 24). Certainly not the Calliope Hummingbird, *Selasphorus calliope*, Martín del Campo's choice, as that species is a winter visitor to northwestern Mexico, and most will at that season lack distinctive ornamentation. Many resident highland species fit better, such as, to suggest but one example, the Violet-crowned Hummingbird, which is gray-brown on the back with a violet crown.

TLE-HUĪTZIL-IN, some species of hummingbird (Trochilidae): "its feathers are glistening, resplendent" (FC, p. 25). Certainly not the Allen's Hummingbird, *Selasphorus sasin*, Martín del Campo's choice, as that species is a winter visitor and is virtually indistinguishable at that season from the more abundant wintering Rufous Hummingbirds. The description could match any number of resident Central Highland species.

CUAPPACH-HUĪTZIL-IN, literally, "tawny hummingbird," likely the Cinnamon Hummingbird, *Amazilia rutila*): "is smoky, dark yellow, tawny" (FC, p. 25). Martín del Campo voted for the Cinnamon Hummingbird, which though restricted to the Pacific slope, is common year around there and is a conspicuous cinnamon color, so I concur.

ECA-HUĪTZIL-IN, literally, "hummingbird of the wind god," the Plain-capped Starthroat (*Heliomaster constantii*) and the White-eared Hummingbird (*Basilinna leucotis*) (Figures 23 and 24):

> It is small and long. Some are ashen, some black. The ashen ones have a black stripe across the eyes. Thus they are painted. And the black ones are

Figures 23 and 24. ECA-HUĪTZIL-IN, White-eared Hummingbird (left, ML432136931, photo by Esteban Matias from Macaulay Library) and Plain-capped Starthroat (right, ML429101601, photo by Yann Muzika from Macaulay Library).

painted on the face with white; they are striped across the eyes with a wind painting <a reference to the wind god, typically illustrated with this distinctive "painted face" (see Figure 36)> (FC, p. 25).

Clearly two species are involved. Martín del Campo opted for the Plain-capped Starthroat, which is a nice fit for the "ashen one." The "black one" is likely the White-eared Hummingbird, a common and conspicuous highlands resident.

TELOLO-HUĪTZIL-IN, one or more hummingbird species (Trochilidae): "It is small and round, small and ashen, small and chalky" (FC, p. 25). This might refer to females of various species.

Finally, one more hummingbird species, this one with its own name:

TOTOZCATLE-TON/TOZCATLE-TON, literally, "little parrot hummingbird," possibly the Broad-tailed Hummingbird (*Selasphorus platycercus*) or the Lucifer Hummingbird (*Calothorax lucifer*): "It is ashen, ash-colored. At the top of its head and the throat, its feathers are flaming, like fire. They glisten, they glow" (FC, p. 25) Martín del Campo voted for the Ruby-throated Hummingbird, *Archilochus colubris*, but that is most unlikely, as that species is a winter visitor, which though common in the highlands during that season, is represented in the great majority by females and non-descript birds of the year. The Broad-tailed Hummingbird shows a brilliant red gorget and is a common resident in the Central Highlands and thus is a better bet, though the Lucifer Hummingbird might fit just as well.

Following the hummingbirds are four more birds, though it is not clear to me why they belong with the parrots and hummingbirds:

YOLLO-TŌTŌ-TL, literally, "heart bird," likely the Rose-throated Becard (*Pachyramphus aglaiae*) (Figure 25):

> lives there in [the province of] Teotlixco, toward the southern sea <the Pacific slope of Chiapas>. It is quite small, the same as a quail. As for its being called *yollotototl*, the people there say thus: that when we die, our hearts turn into [these birds]. And when it speaks, when it sings, it makes its voice pleading; it indeed gladdens one's heart, it consoles one (FC, p. 25).

Martín del Campo suggested that this might be the Rose-breasted Grosbeak (*Pheucticus ludovicianus*) on the strength of the bright rosy throat patch [of the adult male] and its vocal artistry. However, as Martín del Campo noted, this grosbeak is a winter visitor to Mexico. At

Figure 25. YOLLO-TŌTŌ-TL, Rose-throated Becard (painting by Andrew Jackson Grayson in Miller et al., 1957, following p. 62, with permission of the American Ornithological Society).

that season it lacks the colorful adult male plumage and sings rarely. I believe we should look rather to a resident tropical forest species with a similar colorful throat. The Rose-throated Becard seems a more likely candidate.

PŌPOCALES, onomatopoetic, Russet-naped Wood-Rail (*Aramides castaneiventris*) (Figure 26):

> As for its being called *popocales*, it speaks so. Always in the twilight and at dawn, it says *popocales*.... it lives there in [the provinces of] Toztlan [and] Catemahco.... It is the size of a duck, only a little taller.... Its bill is pointed,... chili-red. Its eyes are chili-red. Its head is dark yellow. Its neck, its back, its breast, its tail are ashen.... The legs are chili-red (FC, p. 25–26).

Martín del Campo suggested that this could be a species of rail (Rallidae). In fact, the description, the habitat, and the range all fit the Russet-naped Wood-Rail (*Aramides castaneiventris*) (cf. sound clip by Jessy Lopez Herra, ML435824171).

Figure 26. PŌPOCALES, Russet-naped Wood-Rail (ML435107391, photo by Alex Lamoreaux from Macaulay Library).

The last two species listed in Paragraph Two are both named for their vocalizations. Both are "edible," thus likely hunted for their meat. Martín del Campo did not attempt to identify either. However, I believe both birds are tinamous.

TECUZIL-TŌTŌ-TL is most likely the Thicket Tinamou (*Crypturellus cinnamomeous*).

> it is so named [because] it always so speaks; it indeed pronounces [the sound] *tecucilton, tecucilton*. Its voice is thin. It is the same [size] as a quail, and its feathers are the same…. It lives in…[the provinces of] Teutlixco [and] Toztlan <now in Veracruz state> (FC, p. 26) (cf. sound clip by Vicente Desjardins, ML435021671).

IXMATLA-TŌTŌ-TL, Great Tinamou (*Tinamus major*) (Figure 27).

> Its home is in the forest. It lives there in Anahuac <elsewhere described as "toward the southern sea," i.e., the Pacific Ocean>. It is called *ixmatlatototl* because [its song] is almost like our own speech. When it sings, it says *campa uee*, as if it imitated those who live there …. Its bill is silvery. Its head, its breast, its back, its tail are completely ashen; its feet are ashen (FC, p. 26).

Figure 27. IXMATLA-TŌTŌ-TL, Great Tinamou (ML431413481, photo by Lukas Sekelsky from Macaulay Library).

This fits nicely, as other large tinamous have red feet. Its calls are described as "haunting…. Powerful, eerie, quavering paired whistles …" (Howell & Webb, 1995, p. 89) (cf. sound clip by Adam Dudley, ML432084251).

"Third Paragraph," The Waterfowl

Next up, the "Third Paragraph, which telleth of the waterfowl." The Náhuatl phrase translated as "waterfowl" is "*totome, atlan nemj*," literally, "birds, living among/next to water [Ā-TLĀN]." This paragraph includes 45 entries, nearly a third of all the birds named in the *Codex*. A few may be synonyms, but the diversity of names for waterfowl reflects the rich lacustrine habitat surrounding Tenochtitlán, the island capital of the Aztecs. Seventeen of these entries name ducks (Anatidae); two name grebes (Podicipedidae); five name ralliforms, a rail, the coot, two gallinules, and a crane; eight name egrets, herons, bitterns, an ibis, and a stork. The spoonbill is also considered "waterfowl" but was listed above with the birds of "rich plumage." Also listed are pelicans and a cormorant (Pelecaniformes); four categories of shorebirds, gulls, and terns (Charadriiformes); two swallows (Hirundinidae), and finally, the kingfisher (Alcedinidae). Perhaps by mistake, also included here are a "wild turkey" (*Quauhtotoli*) and an aquatic mammal (*Acujtlachtli*). The facsimile edition clearly shows that *Acujtlachtli* was pasted in at some later date.

One entry is a total puzzle: TEN-ITZ-TLI, literally "obsidian bill," might possibly be a Black Tern (*Chlidonias niger*):

> It flies high always at night there over the lagoon. It is the same size as a dove. Its head is quite small, black. Its breast is somewhat white, somewhat dark. Its back is black; its wings quite small. Its body is all small and round, its tail small, and its legs are like a dove's . . . . it has three bills in all. Its food enters in two places, [though] there is only one throat . . . . It has also two tongues. Its [three] bills are one over the other . . . . And if one captures this [*Tenitztli*], it is a sign that he is about to die . . . . It is a bird of ill omen . . . . The food of the [*Tenitzli*] is water flies, flying ants which fly high (FC, p. 31).

Martín del Campo guessed that this mystery bird might be the Black Skimmer (*Rhychops niger*), given the skimmers extraordinary undercut bill. However, I can see nothing skimmer-like in the description. If I might venture a guess, perhaps the Aztec scribes had in mind the Black Tern, which has been known to hawk insects over marshy lakes at dusk and matches certain of the descriptive details. Given that this bird was feared as an evil omen, it likely was rarely closely observed.

The ducks named nearly cover the Anatid species likely to be encountered on the Valley of Mexico complex of lakes during winter. Martín del Campo had the majority nailed down, but there are several problems with his identifications. Most glaring is his identification of CON-CANAUH-TLI, literally, "pot duck," as the Greater White-fronted Goose (*Anser albifrons*). The *Codex* reports that this "duck" (i.e., a kind of CANAUH-TLI) "is a dweller in the lagoon here, and here it raises its young, builds its nests, lays eggs, sits, hatches its young" (FC, p. 26). However, the goose in question breeds in the high arctic and is at best an occasional winter visitor to the Valley of Mexico. Martín del Campo likewise identifies TLALALACA-TL/Ā-TLATLALACA-TL as this goose, as if it were a synonym. That description is rather different and somewhat more informative: "It is large. The legs are chili-red. [The bill] is ashen, the back is rounded . . . . It has downy feathers; its down is used for capes" (FC, p. 27). No goose is likely to have nested on Lake Texcoco (i.e., "the lagoon"), now or in the past. Of the likely nesting ducks that might fit this description, the Black-bellied Whistling-Duck (*Dendrocygna autumnalis*) is a decent match for TLALALACA-TL. Though whistling-ducks more often nest in coastal lowlands, they have been recorded nesting on highland lakes. If CON-CANAUH-TLI is similar to TLALALACA-TL but distinct, CON-CANAUH-TLI should be the Fulvous Whistling-Duck (*Dendrocygna tricolor*). ZOQUI-CANAUH-TLI "mud duck," might be a synonym.

CANAUH-TLI$_1$ is the general term for "duck," though it is applied a bit more generously than the English term. As is often the case in folk biological taxonomies such as this, the generic term may also do double duty in naming the prototypical species of the generic category, that is, the specific form that most often comes to mind in the broader context. In this case it seems we may have two quite different prototypes. I suggest that we distinguish these prototypical "ducks" as CANAUH-TLI$_2$ and CANAUH-TLI$_3$.

CANAUH-TLI$_2$, a prototypical duck, perhaps the Mexican Duck (*Anas diazi*): "white-breasted; [otherwise] ashen; of average size, not too large. Its breast…, on its belly it is white. It has a wide black bill" (FC, p. 26). This description might serve for the Mexican Duck (*Anas diazi*), in accord with Martín del Campo's translation. Alternatively, this prototypical "duck" might be a catch-all category for the females and/or eclipse plumaged males of many local duck species, which in the winter might not be obviously associated with their respective drakes. Immediately following the second entry labeled CANAUH-TLI is a third entry, also labeled CANAUH-TLI, which is described quite differently.

CANAUH-TLI$_3$, perhaps the Muscovy Duck (*Cairina moschata*): "The head [feathers] are dark green. The head is black. Its head feathers are resplendent, shimmering" (FC, p. 26). Martín del Campo identifies it as the Mallard (*Anas platyrhynchos*), the drake of which displays a metallic green head. However, the true Mallard—as opposed to the "Mexican Mallard," *Anas diazi*, which lacks the green head—is a scarce winter visitor to northern Mexico. I suspect that CANAUH-TLI$_3$ may be the Muscovy Duck, which nests in coastal mangroves and sports metallic greenish-black plumage.

Scattered throughout this paragraph are the following species of ducks, most readily identified, at least with respect to the drakes. We have 12 named categories:

ECA-TŌTŌ-TL, literally, "bird of the wind god," perhaps the Wood Duck (*Aix sponsa*):

> It is called *ecatototl* because its black feathers adorn the face [in the manner of the wind god] …. Its head is quite small; it is crested …. It does not rear its young here; it also migrates …. Many of them come.

Martín del Campo identified this as the Hooded Merganser (*Mergus cuculatus*), but that species is known in Mexico from just a few scattered winter strays. What other crested duck with the distinctive "wind god" face pattern might be a regular winter visitor to the Valley of Mexico? I suspect that the Wood Duck fits the bill (cf. Howell & Webb, 1995, p. 159, 170–171). Note that ECA-HUĪTZIL-IN, literally, "wind god hummingbird," and ECA-TLOH-TLI, literally, "wind god falcon," are both distinguished by a striking face-pattern, resembling images of the god Ehecatl (see Figure 36). "Ehecatl is often portrayed wearing a duckbill mask and a conical hat."

METZ-CANAUH-TLI, literally, "moon duck," the Blue-winged Teal (*Spatula discors*) (Figure 28): "For this reason it is called *metzcanauhtli*: on its face it is decorated with white feathers like the [crescent] moon .... It does not rear its young here; it also migrates" (FC, p. 35). No question here, as the facial marking described is uniquely characteristic of the male of this species.

Figure 28. METZ-CANAUH-TLI, Blue-winged Teal (ML435484041, photo by Brian Stahls from Macaulay Library).

CHIL-CANAUH-TLI, literally, "red duck," the Cinnamon Teal (*Spatula cyanoptera*): "It is named *chilcanauhtli* because its head, breast, back, tail are all like tawny chili: likewise its eyes. Its wings are silvery,... its feet chili-red .... Also it does not rear its young here; it also goes, comes with the others" (FC, p. 38). The details offered fit this species precisely.

YACA-PATLĀHUAC, literally, "wide nose," the Northern Shoveler (*Spatula clypeata*): "It is a duck. It is called *yacapatlauac* because its bill is somewhat long and very wide at the end .... <here they describe how it molts during its winter visit> It also migrates .... Many come here" (FC, p. 38). The description leaves no doubt.

ZŌL-CANAUH-TLI, literally, "quail duck," possibly the Gadwall (*Anas strepera*):

> It is named *çolcanauhtli* because its feathers are all like quail feathers. It is rather large, the same size as a Peru [duck]. White [feathers] are set only

on the point of each wing-bend. Its bill is small and wide; its legs black,…
<here they describe the aquatic plants it likes to eat>. It also comes, it also
migrates with the others…. Many come here (FC, p. 37).

Martín del Campo identified this as the Mallard (*Anas platyrhynchos*). However, nothing
in the description is suggestive, at least, of the male Mallard, with its metallic green head
and conspicuous yellow bill. I vote instead for the Gadwall (*Anas strepera*). Some ambigu-
ity is unavoidable, as the "quail" (ZŌL-IN) of the comparison might call to mind several
different quail species. The Scaled Quail, in particular, would be a good match for the Gadwall's subtle plumage. The white at "the point of each wing-bend" might well refer to the
characteristic white patch on the inner secondaries of this species. Gadwalls are common in
winter throughout central Mexico, unlike the Mallard. The "Peru [duck]" mentioned might
be the Muscovy Duck (*Cairina moschata*), as it had been domesticated by South American
Indigenous communities. It is the size of a small goose. Alternatively, a "Peru duck" might
be a domestic duck—a domesticated form of the Mallard—brought over from Europe by the
conquistadors.

XĀL-CUĀ-NI/XĀL-CANAUH-TLI, literally, "sand eater," the American Wigeon (*Mareca
americana*):

> It is named *xalcuani* because it always eats sand, though sometimes it eats…
> <water plants> [On] its head, its feathers are white on the crown; the sides
> of its head green, resplendent. Its neck is like a quail's; its back ashen….
> Its wings are silvery,…. On both sides [of the bird] are placed tawny feath-
> ers…. It does not rear its young here; it also comes, it also migrates. Many
> come (FC, pp. 36–37).

The details fit the American Wigeon like a glove. The assertion that it "always eats sand" is
puzzling, though wigeon often graze along the shallow margins of ponds, which might have
suggested this interpretation.

TZITZIHUA, the Northern Pintail (*Anas acuta*):

> It is a duck. It is named *tzitziua* because of the feathers growing from its rump;
> among its tail feathers are two very white ones…. Their points are curved;
> they turn upward…. Following down the back of its neck, it is ashen…. it
> also migrates. Neither does it come singly; there are many (FC, p. 36).

Another impeccable description.

Eugene S. Hunn

QUETZAL-TEZOLOC-TON, Green-winged Teal (*Anas crecca*):

> It is a duck called *quetzalteçolocton* because its head is ornamented as if with quetzal feathers. On the crown of its head its feathers are yellow. Its bill is black, small, and wide .... its wings resplendent [green]. But otherwise its wings are all ashen .... It does not rear its young here (FC, p. 34–35).

The contrasting reddish head and green cheek patch is reminiscent of the Resplendent Quetzal's color scheme. The other details, likewise, single out this species.

CUĀ-CŌZ-TLI, literally, "necklace on top," Canvasback (*Aythya valisineria*):

> It is a duck, called *quacoztli* because its head and its neck are tawny to its shoulder. It is as large as a female Peru [duck]. Its eyes are chili-red; its breast white; its back ashen,.... Of its down are made capes. It does not rear its young here, but also migrates (FC, p. 35).

Again, the details provided leave little doubt as to the identity of this species. I would offer only the caveat that the similar and close relative of the Canvasback, the Redhead (*Aythya americana*), might be included here, as both winter regularly to central Mexico. The name is likely derived from CUĀ[I]-TL "head, top ..." and CŌZ-TLI "necklace, collar" (Karttunen, 1983, pp. 43, 58)

TZON-YAYAUHQUI, possibly the Lesser Scaup (*Aythya affinis*) and/or the Ring-necked Duck (*Aythya collaris*):

> It is named *tzonyayauhqui* because its head is very black <TZON-TLI is "head of hair" (Karttunen, 1983, p. 318)>, much like charcoal, reaching to its neck. Its eyes are yellow; its neck, its breast very white; its back dark ashen .... It does not rear its young here; it just comes [and] goes. Many come. They eat what is in the water, [as well as] the sand from the rocks .... (FC, p. 37).

Martín del Campo did not attempt to identify this species. I vote for the Lesser Scaup (*Aythya affinis*), a common winter visitor throughout Mexico otherwise unaccounted for in this otherwise nearly exhaustive account of the ducks now known to winter regularly in the region. Equally plausible is the Ring-necked Duck (*A. collaris*), as suggested by Miller et al. (1957).

AMANACOCHE, the Bufflehead (*Bucephala albeola*):

> It is called *amanacoche* because of its white feathers placed on both sides of its head. It is the size of the <Green-winged> teal…. Its breast is white; its back black;…. It also migrates; it does not rear its young here. Many come (FC, pp. 35–36).

The details leave no doubt that this is the Bufflehead.

Ā-TAPALCA-TL, onomatopoetic, literally, "pot sherds of the water," the Ruddy Duck (*Oxyura jamaicensis*):

> Also [it is called] *yacatexotli*. It is a duck. It comes here in the vanguard, before [other water] birds have come. It is named *atapalcatl* because if it is to rain on the next day, in the evening it begins, and all night [continues], to beat the water [with its wings]. Thus the water folk know that it will rain much when dawn breaks. It is named *yacatexotli* <literally, "has a blue-bill"> because its bill is light blue, small, wide. And near its head … is white. Its head is tawny; its wings, breast, back, tail are all tawny …. It rears its young here; [it has] ten, fifteen, twenty young. Sometimes not all leave; some remain (FC, p. 36).

Once again, a detailed and quite accurate description of this distinctive duck. The name refers to "pottery sherds." The sound of the duck's wings beating the water is like the sound of broken pottery. The Ruddy Duck male "strikes bill against inflated chest in display" and "runs across water to take off" (Howell & Webb, 1995, p. 172), which noises might have inspired the name.

Curiously, the Black-crowned Night-Heron, HUAC-TLI₁, and the Belted Kingfisher, Ā-CHALALAC-TLI, are included with the "ducks": For both, it is said that, "It is a duck." The Common Gallinule (*Gallinula galeata*) is said rather to "belong with the ducks," but is not called a "duck." None of the other waterfowl listed in this paragraph are called "ducks." "Duck" might be considered as a large "folk generic" or named "intermediate" taxon in Berlin's scheme (1992), given that some but not the majority of duck species are named binomially as "X"-CANAUH-TLI.

Two kinds of grebes (Podicipedidae) are named:

Ā-CIH-TLI, Western and Clark's Grebes (*Aechmophorus occidentalis* and *A. clarkii*).

> It is also rare…. It comes when the various [water] birds come….its head is quite small, black, with a pointed, chili-red bill. Its eyes are like fire. It is long-necked…. Its breast very white, its back black,…. Its legs black: they are also somewhat toward its rump, like a duck's legs…. It lives there in the lagoon and is caught in nets. (FC, p. 31) (Figure 41).

The description applies best to Clark's Grebe, as its bill is bright yellow-orange. "Chili-red" might span a range of bright red-orange tones, as do the chilis. The Western Grebe's bill is a duller yellowish. However, both species are resident south to the highlands of Central Mexico and were, until quite recently, considered one species. It is most likely the Aztecs considered them a single species. The name might be interpreted as "water hare" or "water grandmother" (cf. Karttunen, 1983, p. 34), though neither is made explicit in the original account.

YACA-PITZĀHUAC, literally, "slender nose/beak." Eared Grebe (*Podiceps nigricollis*):

> Also [it is called] *nacaztzone*. It is called *yacapitzauac* because its bill is small, pointed, somewhat like a small nail, and it pierces one sharply. And it swims under water; it always feeds under water. And it is named *nacaztzone* because its feathers which are over its ears, inclined toward the back of its neck, are somewhat long, tawny… <perhaps compounded of NACAZ-TLI "ear" and TZON-TLI "head of hair" (Karttunen, 1983, pp. 156, 318)>. Its eyes are like fire, chili-red …. It does not rear its young here; it also migrates (FC, p. 37).

There are recent nesting records in Central Mexico, but it is mostly a winter visitor, as noted.

We next recount the rails, gallinules, and the coot (all Rallidae), species characteristic of marshy pond margins (with the exception of the Russet-naped Wood-Rail, PŌPOCALES, treated above). There are four named categories:

CUA-CHIL-TON/CUA-CHIL-LĬ, literally, "eats chili peppers," the Common Gallinule (*Gallinula galeata*): "It lives on the water; it belongs with the ducks. Its head is chili-red, its bill is pointed. It lives, it is hatched only here, among the reeds" (FC, p. 27). Martín del Campo identifies this as the American Coot (*Fulica americana*). However, the coot's bill is white. The Common Gallinule, close relative of the coot, has a bright yellow bill and a bright red facial shield, so is a better fit.

YACA-CIN-TLI, literally, "corn-cob nose/beak," American Coot (*Fulica americana*): "The Yacacintli is the same as the [Quachilton]" (FC, p. 27). It has been assumed that this is a synonym of CUA-CHIL-TON. However, I believe it is reasonable to interpret this as an assertion that the two are similar but not identical. The Spanish version confirms this: "Hay otra ave semejante a éstas [*cuachilton*], que se llama *yacacintli*" <"There is another bird similar to these [*cuachilton*], that is called *yacacintli*"> (Gárcia Quintana & López Austin, 1988, p. 696). The two species are quite similar in form and behavior, though the gallinule is more retiring.

CUĀ-TĒZCA-TL, literally, "mirror-head," the Purple Gallinule (*Porphyrio martinica*) (Figure 29):

> It likewise is rare. Also it comes when [the other water] birds come …. For this reason it is called [*cuatezcatl*] "mirror-head": on its head is something like a mirror, a round [patch] on the crown of its head …. Its breast, its back are completely light blue; its wings, its tail are likewise light blue …. Its legs are yellow. And when it swims, when it paddles with its feet in the water, it looks like an ember; it goes gleaming, glistening … the appearance of this purple gallinule was [a sign of] war. Whoever captured one looked at himself in the mirror <to foresee his fate in war> (FC, pp. 32–33).

This "mirror" is apparently how the Aztecs interpreted the pale blue facial shield of this exotic-looking bird (Figure 29).

Figure 29. CUĀ-TĒZCA-TL, Purple Gallinule (ML435228071, photo by Annette Teng from Macaulay Library).

COHUIX-IN, onomatopoetic, likely the Aztec Rail (*Rallus tenuirostris*) (Figure 30):

> It is a waterfowl. It is called *couixin* because when it speaks it says *couix, couix*. It is quite small; it is a little larger than a ["paloma"] …. Its bill is chili-red, black at the end, small and cylindrical. Its back, its wings, its tail are all like quail feathers; its breast alone is tawny. It legs are chalky, very long…. This bird also rears its young here; it also comes and it also goes (FC, p. 34).

For some reason, Martín del Campo identified this as the Black-bellied Plover (*Pluvialis squatarola*). However, the Black-bellied Plover is a winter visitor to the coasts and lacks the colors described. I believe the resident Aztec Rail—a secretive but vocal inhabitant of reedy marshes such as surrounded the Aztec capital at the time of the Spanish conquest—fits the description closely, including the vocalizations. Howell and Webb describe this rails calls as "A gruff *ki ka-ka-ka ka-ka-ka….keh-keh-kehrr… kek-kek-kek-kek*" (1995, p. 239) (cf. sound clip by Anuar López, ML432093161).

Figure 30. COHUIX-IN, Aztec Rail (ML430444191, photo by Paula Gatrell from Macaulay Library).

Next up is a distant relation to the rails, a crane (Gruidae):

TOCUIL-COYŌ-TL, the Sandhill Crane (*Grus canadensis*): "The bill is long, like a nail, dart-shaped; the head is chili-red, [the body] ashen, the neck long. It is tall, high, towering. The legs are stringy, very long,… like stilts" (FC, p. 27). Martín del Campo identified it as the

"Brown Crane," now known as the Sandhill. It is now a regular winter visitor to northern Mexico, but rarely ranges south to the Valley of Mexico. It may well have been more common in Aztec times. In any case, it is so distinctive that it would have been hard to overlook.

The Great Blue Heron is often called a "crane" and the Aztecs likewise appreciated their similarity. We list below the herons, egrets, bitterns, and ibis (Ardeiformes), and the Wood Stork (Ciconiiformes). The Roseate Spoonbill, treated above, would have been included here but for its prominence as a source of "rich plumage." The *Codex* lists eight named categories:

Ā-XOQUEN, the Great Blue Heron (*Ardea herodias*): "It resembles the <Sandhill> Crane [in color]; it is ashen, grey. It smells like fish, rotten fish, stinking fish" (FC, p. 28). Martín del Campo concluded that it was the Little Blue Heron (*Egretta caerulea*), apparently because the Spanish version described it as like the crane, but much smaller "*Es de color de las grullas; pero mucho menor*" <"It is the color of the cranes; but much smaller"> (Gárcia Quintana & López Austin, 1988, p. 697). Despite that qualification, I believe that it is much more likely that the Great Blue Heron would have been named than the Little Blue, given that they are equally likely to be encountered in winter in the Central Mexican highlands (Howell & Webb, 1995, pp. 138–140). Note also that the Wood Stork (previously known as the Wood Ibis) is "big, tall, the same as the <Great Blue Heron>" (FC, p. 32).

AZTA-TL, the Snowy Egret (*Egretta thula*): "It is also called *teoaztatl* <sacred egret>. It is white, very white, intensely white.... It is long-necked, stringy, curved. The bill is pointed, long and pointed, black. The legs are very long, stringy, stilt-like, black" (FC, p. 28). It is noteworthy that the Aztecs, the AZ-TECA, derive their ethnonym from their legendary place of origin, AZ-TLAN, which could be interpreted as "place of egrets." It is reasonable to assume that AZTA-TL might include within its extended range the Great Egret (*Egretta alba*), though its bill is yellow.

HUEXO-CANAUH-TLI, literally, "willow duck," possibly the Green Heron (*Butorides virescens*): "Its legs are long, dark green. Its bill is pointed, long and pointed, green .... The head becomes chili-red; its legs become long, rope-like ... stringy" (FC, p. 27). Martín del Campo identified this bird as the Black-crowned Night-Heron. However, the night-heron is better known by the onomatopoetic name HUAC-TLI (see below). The details described fit the Green Heron more closely, plus the Green Heron characteristically forages from stream-side willows. It is called a "duck," as is the night-heron.

HUAC-TLI$_1$, onomatopoetic, the Black-crowned Night-Heron (*Nycticorax nycticorax*):

> It is a duck. It is named *uactli* because its song is like *uactli*: it makes [the sound] *uac, uac*. It is the size of a rooster. The crown of the head is black,

[but] white [feathers] are placed on each side of its head, and at its crown lie three plumes inclined toward the back of its neck—white, well curved. Its bill is black, rounded…. The neck is white … somewhat long. Its breast is also white; on its back it is ashen … its feet yellow…. (FC, p. 39).

Given the detail of this description—which neatly fits the night-heron—it is not likely that a quite different name and description—HUEXO-CANAUH-TLI—should have been used for this species (see above). We should note that HUAC-TLI₂ names the Laughing Falcon (see below in paragraph four), due to the similarity of their characteristic vocalizations.

TOL-COMOC-TLI, onomatopoetic, the American Bittern (*Botaurus lentiginosus*):

Also its name is *atoncuepotli* and *ateponaztli*. It is rather large, the same size as the Castilian chicken, the capon. Its head is dark yellow; it bill yellowish,… cylindrical, about a span <of the hand> in length…. And for this reason is it called *tolcomoctli*: as it sings, it resounds. For this reason it is called *atoncuepotli*: when it sings, it is clearly heard to explode; it is very loud. And it is called *ateponaztli* because it sounds from a distance like a two-toned drum, so loud it is…. This American Bittern always lives here in the reeds; here it raises its young…. When it sings a great deal, <the water folk> know thereby that the rains will come, it will rain much, and there will be many fish—all manner of water life (FC, p. 33).

Compare Sibley's description of the American Bittern's song: "a deep, gulping, pounding *BLOONK-Adoonk* repeated" (2014, p. 107). An iconic bird of marsh land reed beds.

ĀCA-CHICHIC-TLI, onomatopoetic, likely the Least Bittern (*Ixobychus exilis*):

It lives on the water. It is named *acachichictli* because it is as if it sings *achichichic*. And it lives only among the canes, the reeds. And this bird is also an omen for the water folk: always when it sings it is about to dawn…. Its head is quite small; its bill is pointed and small. All of its feathers are yellow, slightly ashen. Its legs are yellow, greenish. It always lives here; it rears its young here (FC, p. 39).

Martín del Campo identified this bird as the Western Grebe. However, that species is better known as Ā-CIH-TLI (see above). Furthermore, the description bears no resemblance to the Western Grebe, neither to the plumage, the soft parts, nor the vocalizations. On the other hand, all these descriptive details fit the Least Bittern.

Ā-CĀCĀLŌ-TL, literally, "water raven," the White-faced Ibis (*Plegadis chihi*): "It is black—a waterfowl, an eater of water life. The legs are long, very long, black; it has a sharp, curved bill" (FC, p. 43). Though described as a "water fowl," this species is listed instead in the fourth paragraph—which "telleth of all the birds [of prey]"—adjacent to the Common Raven (*Corvus corax*). Martín del Campo identified it as the Jabiru (*Jabiru mycteria*), massive older brother to the Wood Stork (see below, CUA-PETLĀHUAC). However, the Jabiru is a rare resident of coastal marshes east of the Isthmus of Tehuantepec. Furthermore, The Jabiru's bill is not "curved," and the body is predominantly white, quite unlike the namesake "raven." I believe the bird named here is the White-faced Ibis, which is resident in the Central Mexican highlands.

CUA-PETLĀHUAC, cf. PETLĀHU(A) "to uncover something," in this case, the head (Karttunen, 1983, p. 192), the Wood Stork (*Mycteria americana*):

> It is a bald-head – big, tall, the same as the <Great> blue heron. Its head is large, like that of our native turkey cock; its head is featherless, bald, bare to the back of the head. The sides of its head are chili-red, reaching to its neck. It is long-necked. Its bill is very thick as well as cylindrical, long, like a bow. Its breast is black. Its back, its wings are completely ashen, except that the wing-bends are very black…. This <Wood Stork> also comes when [water] birds come. It is quite rare (FC, p. 32).

It is noted as an evil omen, foretelling the death of the lords and the outbreak of war. The description fits the Wood Stork (which is the "Wood Ibis" of Martín del Campo) except for the "chili-red" sides of the head. This detail might be due to confusion with the Jabiru. Neither species is likely to have been well known to the Aztecs, as both are coastal species, though the Wood Stork is found all along both coasts.

Two species of pelicans and a cormorant represent the Pelecaniformes; two named categories:

Ā-TŌTO-LIN, literally, "water turkey," the American White Pelican (*Pelecanus erythrorhynchus*):

> It is the ruler, the leader of all the water birds. When the various birds come, this is when it comes; it brings them here at the time of the Feast of Santiago, in the month of July…. And the head of this pelican is rather large, black. Its bill is yellow, round, a span long. Its breast, its back are all white…. Its body is long, very thick…. This pelican does not nest anywhere in the reeds; it always lives there in the middle of the water (FC, pp. 29–30).

Two pages are devoted to this "ruler of all the water birds," describing how the water folk may attempt to capture the bird in hopes of finding a "precious green stone" in its gizzard (Figure 43), though they do so at their peril, as it "sinks people" as they pursue it in their canoes. Though not mentioned in the *Codex*, the other Mexican pelican, the Brown Pelican (*Pelecanus occidentalis*), was also recognized: the *Historia General* notes that, "Hay unas destas [*atotoli*] aves blancas; otras ametaladas" <"There are some of these [*atotoli*] [that are] white; others metallic">(García Quintana & López Austin, 1988, p. 696). I take this to note the contrasting dark plumage of the Brown Pelican, which is strictly coastal.

Ā-COYO-TL, the Neotropic Cormorant (*Phalacrocorax brasilianus*):

> It comes after the pelican [as a waterfowl]. It is also the heart of the water; it also is the leader of the [water] birds .... Its head is as large as the turkey hen's. The bill is pointed, black, quite cylindrical; the outer edge of the tip is yellow. Its breast is rather white. Its back, its wings are all ashen, blackish,.... It is long-bodied. Its legs are thick, not long; they are at its rump, almost at its tail .... It also is rare; it also can sink one (FC, pp. 30–31).

Martín del Campo identified this bird as the "water-turkey," a.k.a., the Anhinga (*Anhinga anhinga*). I favor instead the Neotropic Cormorant. The two species are distant relations and both are primarily coastal residents, though the cormorant breeds across the highlands of northern central Mexico. The description is not precise for either species. The Anhinga's bill is yellow, not black, while the Neotropic Cormorant's dark bill is edged in white, not yellow. The adult cormorant is solid black, not white-breasted, though the young birds may "become whitish" below (Howell & Webb, 1995, p. 129).

The order Charadriiformes includes gulls, terns, stilts, avocets, and a variety of plovers, sandpipers, and a snipe. We have here eight named categories:

PIPITZ-TLI, cf. PIPITZCA "to whinny, shriek, squeak" (Karttunen, 1983, p. 197), the Laughing Gull (*Leucophaeus atricilla*):

> It also lives in the water. Its head is black; its eyes are also black; white [feathers] are set on the eyelids [so that these] appear to be its eyes. It is somewhat long-necked. The throat and breast are white ... on its back, its tail, wings, wing tips, it is black. The tips of both wing-bends are white. Its legs are quite long, chili-red, slender .... Some migrate, some remain and rear their young here. Four are its eggs; only on the ground, on dried mud ... it lays its eggs; not on grass or feathers (FC, p. 39).

This clearly describes the Laughing Gull in high plumage, accurate as well as to its nesting behavior. This is the most common gull species in Mexico and nests along both coasts and rarely in the interior.

PIPIXCAN, cf. PIPIXAHU(I) "to drizzle or snow" (Karttunen, 1983, p. 197), the Franklin's Gull (*Leucophaeus pipixcan*):

> It is white, like a dove; its head is adorned [black and white]. It is a sea dweller, an ocean dweller. It is called *pipixcan* because when it is about to freeze, when the maize is about to be harvested, it comes here (FC, p. 43).

As with the cormorant, this species was exiled from the waterfowl paragraph to paragraph four, which details the birds of prey. This gull is a close relative of the Laughing Gull just described, but is a transient migrant in Mexico, including the highlands, migrating south in late fall, coincident with the maize harvest, as noted. It is intriguing to note that the Latinate species epithet, "*pipixcan*," is borrowed from the Náhuatl. Apparently, the author of the scientific name wished to honor its Aztec affiliation.

XOMO-TL, Royal and/or Elegant Terns (*Sterna maxima, S. elegans*): "It is crested, short-legged, squat. The feet are wide,… blackish. It is a water-dweller, a fish-eater" (FC, p. 27). Martín del Campo averred that this name applied broadly to a variety of similar duck-like species (1940, pp. 393–394). I suspect Martín del Campo is correct that this term encompasses more than one species, but I would argue that the best approximations are the crested terns, notably the Royal and Elegant Terns. Both are regular post-breeding visitors to the Mexican coasts. However, the issue is clouded by additional details presented in the Spanish version of the *Historia General*: "Viven en el agua; también en los montes. Unos dellos son pardos; otros negros; otros blancos; otros cenicientos…. Estos comen peces, y también maiz" <"They live in the water and also in the brush. Some of them are dark, others black, others white, others gray…. They eat fish, and also maize"> (García Quintana & López Austin, 1988, p. 696). These added details missing from the *Codex* might suggest wider application to additional species of tern (Sterninae) or apply instead to the adjacent category, TEZOLOC-TLI (see below).

TEN-ITZ-TLI, literally, "obsidian bill," possibly the Black Tern (*Chlidonias niger*):

> It flies high always at night there over the lagoon [i.e., Lake Texcoco, which is salty]. It is the same size as a ["paloma"]. Its head is quite small, black. Its breast is somewhat white, somewhat dark. Its back is black; its wings quite small. Its body is all small and round…. For this reason it is called "obsidian bill": it has three bills in all. Its food enters in two places, [though] there is

only one throat by which it swallows it. It has two tongues. Its [three] bills are one over the other .... The food of the [obsidian bill] is water flies, flying ants which fly high (FC, p. 31).

This must be the most puzzling account of any bird noted in the *Codex*. Martín del Campo thought that perhaps it was the Black Skimmer (*Rhynchops niger*), which does exhibit a very strangely shaped bill, the lower mandible much longer than the upper, by which it cuts through the water in full flight to catch fish. However, there is nothing else to suggest this equivalence. The skimmer is a coastal species that has rarely strayed to the Central Mexican highlands. It is by no means nocturnal; nor is it insectivorous; also, it is quite large and long winged. I have been perplexed over this and can only guess that if such a bird actually existed, it might be the Black Tern (*Chlidonias niger*), which is a common migrant across Mexico, including across the highlands, and characteristically hawks insects in buoyant flight. Furthermore, "... if one captures this <bird>, it is a sign that he is about to die .... his house will be destroyed; his house folk will perish" (FC, p. 31). It seems unlikely the Water Folk would have gone out of their way to capture one to have a closer look.

TEZOLOC-TLI/TEZOLOC-TON, various shorebird species (*Charadrius* spp., *Calidris* spp., et al.): "... is small, one which whirrs [as it flies]" (FC, p. 27). Martín del Campo averred that this name, as well as XOMO-TL (see above), applied broadly to a variety of similar species (1940, pp. 393–394). The Spanish version of the *Historia General* in describing XOMO-TL noted that, "They live in the water and also in the brush. Some of them are dark, others black, others white, others gray .... They eat fish, and also maize" <author's translation>. This is quite at odds with what was noted for XOMO-TL in the *Codex*. However, the elaborations noted above, if they applied instead to TEZOLOC-TLI, might be readily understood as descriptive of the many and diverse species of shorebirds that would have visited the lakes and fields of the Valley of Mexico, after nesting, in migration, and/or during the winter. By my rough count, at least 16 species of plovers and sandpipers might be expected in the central Mexican highlands. Flocks of such birds might settle on lake shores and open fields, flushing ("whirring") at the slightest hint of danger.

Ā-TZITZICUILO-TL, literally, "small stick of the water," the Black-necked Stilt (*Himanto-pus mexicanus*) conflated with Wilson's and Red-necked Phalaropes (*Phalaropus tricolor* and *P. lobatus*): We have two descriptions, the Náhuatl of the *Florentine Codex* and the Spanish of the *Historia General*. They agree on various points, but in combination pose a puzzle. First, the Náhuatl: "It is round-backed. The bill is long, needle-like, pointed, ... black. The legs are long, very long, stilt-like, broom-like, slender. Its dwelling place is [the province of] Ana-huac. It is white-breasted ...." (FC, p. 28). Martín del Campo identified this as the Red-necked (=Northern) Phalarope (*Phalaropus lobatus*). However, that species breeds on arctic tundra ponds and migrates along the coasts. It is, thus, most unlikely to have been familiar to the Aztec scribes. Given that it is said that "its dwelling place is Anahuac," one might assume that it was a resident, rather than a migrant. The best fit for these details would be the Black-necked Stilt, which is a "C[ommon] to F[airly common] resident (S[ea] L[evel] – 1500 m[eters]" throughout Mexico (Howell & Webb, 1995, p. 261). The Spanish version of the *Historia General* adds details that complicate the identification: "*atzitzicuilotl*" are described as "*cenicientos*," that is, "ashy," and it is said that they visit the "laguna de México" with the rains, and that flocks of these birds later disappear into the ocean where they turn into fish (García Quintana & Lopez Austin, 1988, p. 696). Black-necked Stilts might well disperse "with the rains" to the Valley of Mexico lakes after nesting at lower elevations. However, they are black above, not "ashy," and disappearing beneath the waves at the seashore hardly fits. Perhaps the scribes confounded the resident Black-necked Stilt with the migratory Wilson's Phalarope, which, unlike the Red-necked Phalarope, migrates throughout the central highlands. It is a better fit for the Náhuatl descriptive details than the Red-necked, in that it has a longer, thinner bill and is more likely to be seen foraging on shore on its long legs, details shared with the stilt but contrasting with the Red-necked Phalarope, which characteristically swims in tight circles on the surface chasing insects, rarely exposing its legs. Wilson's Phalaropes in winter are "ashy" above, white below, thus matching the details of the Spanish account somewhat better than the Stilt. Finally, flocks of Red-necked Phalaropes might have contributed to this compound description, as they may be abundant offshore in migration and can in fact appear to vanish beneath the waves like flecks of sea foam. We are left with a descriptive quilt that might have been inspired by all three species. Phalaropes are known in the contemporary Mexican vernacular as "chichicuilotes," clearly derivative of the Náhuatl. León García Garagarza muses,

> Estos chichicuilotes, o falaropos, afortunadamente son todavía abundantes, así es que la próxima vez que los vean en la playa, consideren que los ancestros mexicanos estaban convencidos de que, cuando ya no están ahí, es porque se han echado en masa al agua y se han convertido en peces, en verdaderas sírenas pajareras ["These chichicuilotes, or phalaropes, fortunately are still abundant today, so the next time you see them at the beach, take note that the ancestral Mexicans were convinced that when they were no longer present they had entered the sea all together and were transformed into fishes, avian mermaids in truth", my translation] (L.G. Garagarza, personal communication, August 16, 2022).

ICXI-XOXŌUHQUI, the American Avocet (*Recurvirostra americana*):

> It is a waterfowl. It is named *icxixoxouhqui* because its legs are green <Sibley describes the legs as "pale bluish" (2014, p. 165), a tint likely covered by the Náhuatl color term>. The bill is small and cylindrical, small and

slender, black, curved upwards. Its head is quite small, white; it is rather long-necked. Its breast, its back are white;…. Quite small are its wings; the upper surfaces are black, and the under surfaces quite white;…. And when it has shed, its head and neck are almost chili-red,…. It raises its young here,… when the rains come (FC, p. 34).

There can be no doubt that this is the American Avocet, with its recurved bill and reddish breeding plumage.

Ā-ZŌL-IN, literally, "water quail," the Wilson's Snipe (*Gallinago delicata*): "Also it is called, çoquiaçolin <"mud quail">. It is long-billed, long-legged. It is varicolored like a quail…. Its home is in the water, among the reeds" (FC, p. 28). There is no doubt about this identification. It is a common winter visitor throughout Mexico, including the highlands.

Concluding this third paragraph which "telleth of the waterfowl" are three quite different species, though all three are closely associated with aquatic habitats:

Ā-CHALALAC-TLI, onomatopoetic, the Belted Kingfisher (*Ceryle alcyon*):

It is a duck [i.e., *canauhtli*]. It is named *achalalactli* because it sings thus: *cha, chacha, chuchu, chala chala chala* <cf. sound clip by Liza Verkony, ML436910561>. It is the size of a [shorebird]. But this bird does not live in the brackish lagoon; rather it lives in the fresh water. It frequents the crags. Also it does not settle upon the water but always goes to alight upon the tops of willows, on treetops. And when it wishes to feed, from there it descends, suddenly dives into the water, it takes what it hunts, perhaps a fish, perhaps a frog. And when it has gone to take it, then it calmly goes to the treetop there to eat it …. Its head is crested; ashen are its feathers …. Its bill is black, small, pointed, cylindrical …. Its breast is white, its back dark gray,… It always lives here; [there are] not very many; they are somewhat rare. It is not known where they rear their young (FC, p. 38).

Though it seems curious to call the kingfisher "a duck," the descriptive details are impeccable.

Ā-CUICUIYALO-TL/A-CUECUEYALO-TL, the Cliff Swallow (*Petrochelidon pyrrhonota*) (Figure 31): "It nests in the crags and house roofs, house fronts. It is a mud-nest builder which plasters with its bill…. It is [dark] ashen, like [CUĪCUĪTZCA-TL]" (FC, p. 28). This swallow and the next are the two species of the central Mexican highlands that build mud nests (Figure 31).

Figure 31. Ā-CUICUIYALO-TL, Cliff Swallows nesting (Dibble & Anderson, 1963, Figure 81, after Paso y Troncoso).

CUĪCUĪTZCA-TL, cf. CUĪCA-TL "song" (Karttunen, 1983, p. 71), the Barn Swallow (*Hirundo rustica*):

> It is small and black, with small, pointed bill, with small, short legs. It is charcoal colored,…. It is a warbler, a crier, a constant warbler, an awakener of the sleeping. It is a builder of mud nests in house roofs, in house fronts. It is a traveller, a disappearer; later it comes, in [the month of] Atemoztli …. It cleans itself, beautifies itself; it hurls itself into the water, it bathes itself (FC, p. 28).

It is noteworthy that the Aztec scribes were not content to provide just the barest details for a favorite bird but to wax lyrical.

"Fourth Paragraph," The Birds of Prey

Next, we consider the "Fourth Paragraph, which telleth of all the birds [of prey] [IXQUICHTIN TOTO-ME, TLAHUIITEQUI-NI]." "The different eagles mentioned inhabit inaccessible places, among the crags; [there] they nest, lay eggs, hatch their young <There follows an account of how men captured the young of these birds>" (FC, p. 42). This section is roughly evenly divided between two diverse generic categories, CUĀUH-TLI, which in its generic sense includes a variety of eagles and large hawks, and TLOH-TLI, which also has the generic sense of "falcon" but may also include a few smaller hawks. The two categories are not perfectly distinct, as we have two specific categories that seem to bridge the divide: CUĀUH-

TLOH-TLI, literally, "eagle-falcon," and TLOH-CUĀUH-TLI, which should mean literally, "falcon-eagle." Many of the specific kinds of "eagle" and "falcon" cannot be positively identified given the great diversity of such species that the Aztecs might have encountered and the often-confusing diversity of plumages a given species might exhibit, plus the difficulty of observing such birds in flight without the aid of binoculars. It is often better to admit that no positive equivalence to a particular raptor species is justified by the published descriptions. There appears to be an excess of named categories with which to equate the likely raptor species due to the unusual number of apparent synonyms. Also treated in this chapter is the Common Raven, White-faced Ibis, Franklin's Gull, and the Loggerhead Shrike. I discussed the ibis and gull above with the waterfowl.

CUĀUH-TLI$_1$, "eagle" (Accipitridae in part): This term serves as the head term in several binomial names for a variety of large raptors, as noted below. It may also name the Golden Eagle unmodified or with the attributive "obsidian."

CUĀUH-TLI$_2$/ITZ-CUĀUH-TLI, "obsidian eagle," Golden Eagle (*Aquila chrysaetos*) (Figure 32):

> The eagle is yellow-billed…. The bill is thick, curved, humped, hard. The legs are yellow…. They are thick. The claws are curved, hooked. The eyes are like coals of fire. It is large, big. On its head, and neck, and on its wings, on its wing-bends, and on its back lie feathers called *quauhtapalcatl*. <Next follows the enumeration of seven additional named varieties of feathers.> …. The eagle is fearless, a brave one;… it can gaze into, it can face the sun …. It is ashen, brown. It beats its wings,… it constantly grooms itself (FC, p. 40).

Figure 32. CUĀUH-TLI$_2$, Golden Eagle (Dibble & Anderson, 1963, Figure 114, after Paso y Troncoso).

This description would seem to single out the Golden Eagle, though the name, unmodified, might be extended to a number of larger hawk species …

> It is called *itzquauhtli* because the feathers of its breast, of its back are very beautiful; they glisten as if blotched with gold, and they are called *quauhxilotl*. Its wings, its tail are blotched with white;…. And it is called *itzquauhtli* because it is a great bird of prey. It preys on, it slays the deer, the wild beasts…. It can slay very thick snakes, and can kill whatever kind of bird flies in the air. It carries them off wherever it wishes to go to eat them (FC, p. 41).

MIX-CŌĀ-CUĀUH-TLI, cf. MIX-TLI "cloud," CŌĀ-TL "snake" (Karttunen, 1983, pp. 36, 149), Ornate Hawk-Eagle (*Spizaetus ornatus*) (Figure 33):

> … is dark, its face is adorned…. It is not very large; average in size, somewhat the same as the turkey hen living here. It is named *mixcoaquauhtli* because at the back of its head are its feathers, paired feathers forming its head pendant. It is white across the eyes, joined, touching the black; so is the face adorned…. It lives everywhere and is also a bird of prey (FC, p. 41, also Figure 122).

Martín del Campo identified this raptor as the "crab-hawk," now known as the Common Black Hawk (*Buteogallus anthracinus*). However, that species lacks the crest feathers that set this species apart. The only raptor that fits this description that is resident in central Mexico is the Ornate Hawk-Eagle. The details of the facial "adornment" likewise better match the hawk-eagle than the black-hawk. The Harpy Eagle (*Harpia harpyja*) is also crested, but is larger than the Golden Eagle and, at least in recent years, restricted to the Atlantic lowlands north to southern Veracruz. If the Aztecs knew of such a bird it seems highly likely they would have named it here.

Figure 33. MIX-CŌĀ-CUĀUH-TLI, Ornate Hawk-Eagle (ML434456861, photo by Sergio Andrés Cuellar Ramírez from Macaulay Library).

IZTAC-CUĀUH-TLI, literally, "white eagle/hawk," perhaps the White-tailed Hawk (*Buteo albicaudatus*): "It has scant, ashen [feathers]; it lacks down; it is very chalky. The bill is yellow, the legs are yellow" (FC, p. 40, see also Figure 116). Two species might fit this description, brief as it is. The White Hawk (*Leucopternis albicollis*) is striking, nearly immaculate white. It is resident in the Atlantic lowlands north to southern Veracruz state. The White-tailed Hawk (*Buteo albicaudatus*) is white below and on the tail, but otherwise gray. It is more widespread than the White Hawk and may occur at mid-elevations across central Mexico. The Ferruginous Hawk (*Buteo regalis*) is likewise largely white below and on the tail. It is a winter visitor to northern Mexico. One cannot be certain, but I suspect the White-tailed Hawk is the most likely candidate.

YOHUAL-CUĀUH-TLI, literally, "nocturnal eagle," possibly the Northern Potoo (*Nyctibius jamaicensis*): "It is the same as the white eagle. It is called 'nocturnal eagle' because it appears nowhere much by day, but at night it eats, it preys on [its victims]" (FC, p. 40, see also Figure 117). I must admit that the identity of this bird is a major puzzle. I can think of no owl (Strigiformes) or nightjar (Caprimulgidae) that could be compared to a large hawk, and no hawk that is at all nocturnal. The Northern Potoo with a wingspan of 0.91–1.02 m, double that of the Lesser Nighthawk (*Chodeiles acutipennis*), is a possible candidate, as the White-tailed Hawk's wing span is 1.24–1.37 m (Howell & Webb, 1995, pp. 201, 383).

TLACO-CUĀUH-TLI, literally, "middle eagle/hawk," perhaps the Northern Harrier (*Circus cyaneus*): "Also it is called *chienquauhtli* [CHIYAN-CUAUH-TLI, literally 'chia eagle']. It resembles a <COHUIX-IN, Aztec Rail>. It has a yellow beak, it has yellow legs" (FC, p. 40, see also Figure 118). Martín del Campo identified this species as the Northern Harrier (then known as "Marsh Hawk"). How that species might resemble an Aztec Rail <or a "Black-bellied Plover," as Martín del Campo suggested> is a puzzle. Note also that the Northern Harrier was also attributed to TLOH-CUĀUH-TLI, literally, "falcon-eagle," perhaps in error (see below). Nevertheless, this species would have been a common and conspicuous migrant and winter visitor to the Valley of Mexico wetlands and thus, likely, named.

Ā-CUĀUH-TLI, literally, "water eagle/hawk," possibly the Crane Hawk (*Geranospiza caerulescens*): "It is of average size, not very large. The water is its habitat. It preys upon, it eats the waterfowl" (FC, p. 41, see also Figure 119). One might presume this to be the Osprey, but that species is named Ā-ITZ-CUĀUH-TLI and is carefully described below. What other raptor might fit this description? I vote for the Crane Hawk, given its habitat preferences: "Mangroves, swampy woodland, thorn forest, semiopen areas…mostly near water…. (S[ea]L[evel]-1500 m) on both slopes [from northern Mexico south]" (Howell & Webb, 1995, pp. 189–190).

Ā-ITZ-CUĀUH-TLI, literally, "water Golden Eagle," Osprey (*Pandion haliaetus*):

> It is like the Golden Eagle …. It is named *aitzquauhtli* because when it goes flying high, if it wishes to eat, from there it streaks down. When it descends, it goes whirring, it suddenly dives into the water; it seizes whatever it wishes to eat, perhaps a fish … (FC, p. 41, see also Figure 121).

This could only be the Osprey, which nests along both Mexican coasts and ranges in winter throughout the central Mexican highlands.

COZCA-CUĀUH-TLI, literally, "yellow eagle," perhaps the Crested Caracara (*Caracara plancus*): "It is smoky. The wings are curved, resembling an eagle's. The bill is curved" (FC, p. 42, see also Figure 123). Martín del Campo identified this bird as the King Vulture (*Sarcoramphus papa*). However, this seems most unlikely. The King Vulture would more likely be treated as a relative of or kind of TZOPĪLŌ-TL. It is of outstanding size, nearly the wing-span of the Golden Eagle, and with highly distinctive plumage. It is rare away from its lowland tropical forest habitat, though with a somewhat wider historic distribution (Howell & Webb, 1995, p. 176). I am inclined to equate this with the caracara, which is a widespread and quite familiar raptor otherwise unrecognized.

Next the *Codex* lists seven "birds of prey" that are neither named "eagle" nor "falcon": the Laughing Falcon, the vulture, the owl, the "grass owl," the Common Raven, the Jabiru (*sic.*), and the Franklin's Gull. These birds hunt or scavenge and are thus seen to be related to the true raptors.

HUAC-TLI$_2$, onomatopoetic, Laughing Falcon (*Herpetotheres cachinnans*):

> It resembles the <Crested Caracara, not the "King Vulture," as Martín del Campo suggested>. It sings in this manner: sometimes it laughs like some man; like a man speaking it can pronounce these words: *yeccan, yeccan, yeccan.* When it laughs, it says *ha ha ha ha ha, ha hay, ha hay, hay hay, ay.* Especially when it finds its food it really laughs <cf. sound clip by Adam Betuel, ML436679641> (FC, p. 42).

There can be no doubt that this is the Laughing Falcon, a conspicuous and common resident at low to mid-elevations throughout Mexico. Its name is a homonym of the Black-crowned Night-Heron (see above), a coincidence of the resemblance between their characteristic vocalizations.

TZOPĪLŌ-TL, Black and Turkey Vultures (*Coragyps atratus* and *Cathartes aura*): "It is black, dirty black, chili-red headed, chalky-legged. All its food is what has died—stinking, filthy"

(FC, p. 42). It seems the Aztecs had the same opinion of the vultures as we do. The "red-headed" vulture is, of course, the Turkey Vulture, and that is clearly the bird illustrated in the *Codex* (Figure 125). However, The Spanish version is more suggestive of the Black Vulture: "Son negras …. Andan en bandas,…. Andan cerca los pueblos" <"They are black …. Go around in groups …. Go around in towns"> (García Quintana & López Austin, 1988, p. 706). The contemporary Mexican vernacular term, *zopilote*, refers in particular to the Black Vulture (Martín del Campo, 1940, p. 402). Apparently, the Aztecs did not sharply distinguish these two common vultures.

TECOLŌ-TL, onomatopoetic, Great Horned Owl (*Bubo virginianus*):

> It is round, like a ball …. The eyes are like spindle whorls; shiny. It has horns
> of feathers. The head is ball-like, round; the feathers thick, heavy. It is blind-
> ed during the day. It is born in crags, in trees. It feeds by night, because it
> sees especially well in the dark. It has a deep voice when it hoots; it says
> *tecolol tecolo, o, o* (FC, p. 42).

The "horns of feathers" and the vocalizations single out the Great Horned Owl, the most powerful and widespread of Mexican owls. It is likely that other owl species would also be called *tecolotl*, though none—with but three exceptions, the Striped Owl/Short-eared Owl, the Burrowing Owl, and the Barn Owl—appear to have been distinguished by name by the Aztecs. "Tecolote" is contemporary Mexican Spanish for "owls" in general, derived from the Náhuatl (Schoenhals, 1988, pp. 475–476).

ZACA-TECOLŌ-TL, literally, "grass owl", possibly, the Striped Owl and/or the Short-eared Owl (*Asio clamator* and/or *A. flammeus*): "It is small, blotched like a quail, only it is also like the *tecolotl*. It is called çacateolotl because it is born in the grasslands" (FC, p. 42). Martín del Campo identified this as the Burrowing Owl (*Athene canicularia*), which seems quite reasonable, except that an apparently different bird, TLĀL-CHICUA-TLI, is described in fine detail, leaving no doubt that that bird is the Burrowing Owl (see below). Assuming that these two names contrast, I am inclined to identify this "grass owl" as either the resident, lowland Striped Owl or the winter visitor Short-eared Owl, both which characteristically inhabit savannahs and marshlands (Howell & Webb, 1995, pp. 368–369).

CĀCĀLŌ-TL, onomatopoetic, the Common Raven (*Corvus corax*):

> Also it is called *calli* and *cacalli*. It is really black, really charcoal-colored, a
> well-textured black: very black…. It is an eater of ripe maize ears, of tunas

<prickly-pear cactus fruits>, of mice, of flesh. It stores things... it piles the ripe maize ears up within [the hollows of] trees (FC, p. 43, see Figure 128).

Certainly, this could be none other than the Common Raven, often mistakenly called "cuervo" or "crow." However, crows are unknown south of northern Mexico, while ravens range throughout Mexico.

Ā-CĀCĀLŌ-TL, literally, "water raven," the White-faced Ibis, is dealt with above with the waterfowl.

PIPIXCAN, Franklin's Gull (*Leucophaeus pipixcan*). This bird is considered above with the waterfowl.

Following this diversion we return to the true raptors, falcons, plus possibly an *Accipiter* and a harrier. We have more names here than we know what to do with:

TLOH-TLI$_1$/TO-TLI$_1$, falcon in general (*Falco* species) and other small hawks (Accipitridae in part). This term serves as the head for the binomial names of seven raptor species. Unmodified, it may also name the prototypical falcon, which for the Aztecs appears to have been the Prairie Falcon (see below).

TLOH-TLI$_2$/TO-TLI$_2$, Prairie Falcon (*Falco mexicanus*): "Also its name is *totli*. It is ashen. It is a hunter, a bird of prey; a whirrer; a flesh-eater. Its bill is pointed, curved; green, dark green. Its legs are dark green" (FC, p. 43). The Prairie Falcon today occurs south to the northern Mexican highlands, breeding south to Durango and San Luis Potosi and wintering south as far as Jalisco and Hidalgo, just north of the Valley of Mexico. In Aztec times, perhaps, it was a regular visitor farther south. In any case, of the falcons now known to frequent the highlands of central Mexico, this species and the Peregrine are the most "charismatic." As I argue below, the Peregrine is likely named CUĀUH-TLOH-TLI, contrasting with the generic prototype.

TLOH-CUĀUH-TLI/TLĀCO-TLO-TLI, literally, "falcon-eagle" or "staff falcon," likely the Northern Goshawk (*Accipiter gentilis*): "It is also called *tlacotlotli*. It is large, ashen. It hunts rabbits" (FC, p. 43). Martín del Campo identified this species as the Northern Harrier (now known as "Marsh Hawk"), though he had previously assigned that species to another category, TLACO-CUĀUH-TLI, a type of "eagle." I would like to suggest an alternative identification, as the Northern Goshawk. The Northern Harrier would not hunt rabbits, as it forages for mice over marsh and grasslands and lacks the power. The goshawk is a resident of the forested highlands of north-western Mexico south to Guerrero. It is a powerful raptor that "Hunts... for grouse, rabbits, and squirrels, launching explosive attack once prey is sighted"

(Sibley, 2014, p. 129). This hunting style seems more falcon-like than the leisurely pursuit of the harrier. The Spanish version in the *Historia General* supports this interpretation, equating this with "*azores* [that is, the goshawk] *como los de España*" (García Quintana & López Austin, 1988, p. 402).

CUĀUH-TLOH-TLI/TLO-CUĀUH-TLI, literally, "eagle-falcon"/"falcon-eagle," the Peregrine Falcon (*Falco peregrinus*) (Figure 34):

> Likewise [it is called] *tloquauhtli*. The hen is somewhat large, and the cock somewhat small. The hen is a great hunter. It is called a falcon <halcón in Spanish>. It has a yellow bill; its feathers are all dark grey; there are twelve [feathers] on its tail. Its legs are yellow. When it hunts, [it does so] only with its talons. When it goes flying over birds,... it only tries to seize them with its talons.... And if <the *quauhtlotli*> succeeds in catching one, it at once clutches [the victim] by the breast; then it pierces its throat. It drinks its blood; consumes it all... first it plucks out the bird's feathers.... It brings forth its young in inaccessible places; it nests in the openings of the crags. It has only two young.... And this falcon gives life to Uitzilopochtli <"Hummingbird-of-the-Left," patron warrior god of the Aztec people> because,

Figure 34. CUĀUH-TLOH-TLI, Peregrine Falcon (Dibble & Anderson, 1963, Figure 133, after Paso y Troncoso).

68

they said, these falcons … when they eat … as it were give drink to the sun (FC, pp. 43–44).

The Aztecs also capture these birds alive. This is a description worthy of the finest of our natural history writers.

COZ-TLOH-TLI, literally, "yellow falcon," the American Kestrel (*Falco sparverius*): "It is quite small. And the cock is called <in Spanish> *tuercuello*. It is named "yellow falcon" because its feathers are yellow" (FC, p. 44). The Spanish version in the *Historia General* is a bit more explicit: "Ay también cernjcales como los de españa, y la color dellos es como la color de los de españa" <"There are also kestrels like those of Spain, and their color is like the color of those of Spain"> (García Quintana & López Austin, 1988, p. 707). This can only be the American Kestrel, which closely resembles the European Kestrel or *cernícalo* (*Falco tinnunculus*).

ECA-TLOH-TLI, literally, "wind god falcon," the Aplomado Falcon (*Falco femoralis*) (Figures 35 and 36): "It is the same as the falcon <Halcón, from Spanish>. It is named *ecatlotli* because its face is shot across with white. Its feathers are somewhat dark. It hunts in the same way as has been told [of the Peregrine Falcon, above]" (FC, p. 44). This must be the Aplomado Falcon (*Falco femoralis*), now quite rare in northern Mexico but more widespread in the highlands of Central Mexio historically (Howell & Webb, 1995, pp. 216–217). This falcon shows the distinctive contrasting white eye-line through the dark crown suggestive of the Aztec wind god, EHĒCA-TL (see Figure 36 below), a naming pattern found also with a hummingbird and a duck (see above).

Figure 35. ECA-TLOH-TLI, Aplomado Falcon (ML432785111, photo by Dave Beeke from Macaulay Library).

Figure 36. EHĒCA-TL, the wind god.

Eugene S. Hunn

ĀYAUH-TLOH-TLI, literally, *"cloud falcon,"* likely another name for the Peregrine Falcon (*Falco peregrinus*):

> Its name is [also] *moralo* [apparently from Spanish, meaning unknown]. This one is named *ayauhtlotli* because it hunts and strikes in the clouds. All said of the falcon <"halcón," the Peregrine Falcon> is the same … except that its feathers are quite ashen, like the <Sandhill Crane's> feathers (FC, p. 44).

As plumage coloration varies by sex, age, and subspecies, this is likely an alternative name for the Peregrine Falcon, well-known for hunting from the heights.

IZTAC-TLOH-TLI, perhaps another name for the Prairie Falcon (*Falco mexicanus*):

> Its name is [also] *sacre* <from Spanish for the Saker Falcon, *Falco cherrug*, which closely resembles the Prairie Falcon>. It is large and tall, very strong. It does not hunt ducks much; it wars upon hares, rabbits, turkeys, and chickens. It is called "white falcon" because its feathers are pale striped with white. It always hunts [its prey] by day,… Its legs are yellow (FC, p. 44).

ITZ-TLOH-TLI/ĀCA-TLOH-TLI/TLETLEUH-TZIN, literally, "obsidian falcon," "reed falcon," and "little Merlin," perhaps the Bat Falcon (*Falco rufigularis*):

> Or else [it is called] "reed falcon." Its name is [also] *gavilán* <"hawk," from Spanish>. It is named "reed falcon" or "obsidian falcon" because its bill is quite long and narrow, like an obsidian point. Its feathers are quite smoky, dark. And its tail is somewhat long, white mingled [with black] …. Its name is also *tletleuhtzin*. It is small, bold; a whirrer; a bird of prey (FC, pp. 44–45).

The synonym here, TLETLEUH-TZIN, with the honorific/diminuitive suffix -TZIN, appears to contrast with one of the Merlin's alternative names, TLETLEUH-TON (see below), though TŌN-TLI likewise suggests small size (Karttunen, 1983, pp. 247, 314). The description of the bill is odd, as both the Bat Falcon's and the Merlin's bills are typical falcon shape. If this is in fact the Bat Falcon, YOHUAL-TLOH-TLI, literally, "nocturnal falcon," would be yet another synonym. There seems a proliferation of names for falcons with perhaps some "over-differentiation" (Berlin, 1992).

YOHUAL-TLOH-TLI, literally, "nocturnal falcon," likely the Bat Falcon (*Falco rufigularis*): "It is the same as a falcon <TLOH-TLI>. It is named *youaltlotli* because it sees in the dark: it can hunt, it can strike its prey" (FC, p. 45). As there are no truly nocturnal falcons, Martín del Campo suggested this might be the Lesser Nighthawk (*Chordeiles acutipennis* of the Caprimulgidae), which, though nocturnal, hawks insects and would never "strike its prey." The Bat Falcon, as its name implies, hunts its characteristic prey at dusk, thus could have been seen as a "nocturnal falcon." If so, ITZ-TLOH-TLI, ĀCA-TLOH-TLI, and TLETLEUH-TZIN, literally, "obsidian falcon," "reed falcon," and "small Merlin" would be synonyms.

NECUILIC-TLI/NECUILOC-TLI/ECA-CHICHIN-QUI/CENOTZ-QUI/TLETLEUH-TON, the Merlin (*Falco columbarius*):

> —or *neculotli*, or *e[he]cachichinqui* [literally, "air-sucker"]; and they name it *cenotzqui* <literally, "calls the frost"> and *tletleuhton* <literally, "small fire">. It is of average size. The bill is pointed, small and pointed. It is an eater of mice, of lizards, of çacacilin <unidentified birds, though perhaps "sparrows," ZACA-TLA-TLI/ZACA-TLA-TON, literally "grass eaters," see below>. It is an air-sucker. It is [spotted] yellow and black. When it has eaten, then it sucks in air; it is said that thus it gets water. And from the wind it knows when the frost is about to come. Then it begins to sing (FC, p. 45).

Martín del Campo identified this multiply-named species as the Merlin. As the Merlin is a common winter visitor, the association with frost seems appropriate (Martín del Campo 1940:403). In this instance, the prefixed modifier ECA- refers to its association with winds, not because of any resemblance to the image of the wind god, EHECA-TL, as was true of the duck, hummingbird, and falcon similarly named. Note also that what I believe to be the Bat Falcon may be called TLETLEUH-TZIN, the suffix –TZIN conveying an honorific or diminuitive sense (Karttunen, 1983, p. 314).

To recap the falcon nomenclature: there are two names that I believe apply to the Prairie Falcon, IZTAC-TLOH-TLI "white falcon," plus the generic TLOH-TLI$_2$ "falcon," for which it is the prototype. There are three names for the Peregrine, CUĀUH-TLOH-TLI "eagle-falcon," TLO-CUĀUH-TLI "falcon-eagle," and ĀYAUH-TLOH-TLI "cloud falcon." There are four names for the Bat Falcon, ITZ-TLOH-TLI "obsidian falcon," ĀCA-TLOH-TLI "reed falcon," TLETLEUH-TZIN "small Merlin," and YOHUAL-TLOH-TLI "nocturnal falcon." Finally, we have four synonyms for the Merlin, NECUILIC-TLI, ECA-CHICHIN-QUI "wind sucker," CENOTZ-QUI "calls the frost," and TLETLEUH-TON "small fire." A multiplicity of names often indicates exceptional cultural value. It is also interesting to note that two species serve to bridge the distinction between "eagle" (CUĀUH-TLI) and "falcon" (TLOH-TLI). The Peregrine Falcon may be known as CUĀUH-TLOH-TLI "eagle-falcon," while the bird I have identified as the Northern Goshawk may be known as TLOH-CUĀUH-TLI "falcon-eagle." The Peregrine and the Northern Goshawk seem best suited for this role linking the two basic types of raptors.

Seemingly, as an afterthought, the *Codex* adds the shrike to this fourth paragraph, "which telleth of all the birds [of prey]" It is an unusual "song bird" that is a most effective hunter of small prey.

TETZOMPAH/TETZOMPAH MĀMANA, Loggerhead Shrike (*Lanius ludovicianus*) (Figure 37):

> Its wings are mingled white [and black]. Its bill is pointed, like a metal awl.
> It is called *tetzompa* because, when it has fed, when it is satiated, it impales
> its catch – mice, lizards – on trees [and] on maguey leaves <the spiny leaves
> of the agaves, also known as "century plants"> (FC, p. 45) (Figure 37).

The shrike is famous for this prey cacheing behavior. It is a common resident throughout Mexico south to the Isthmus of Tehuántepec.

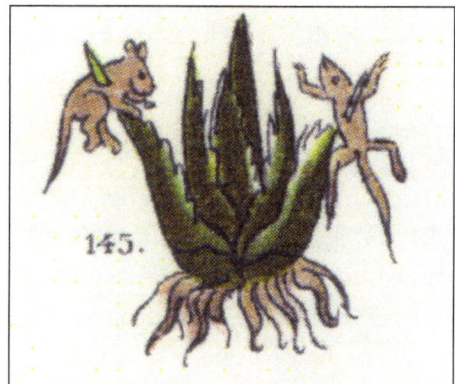

Figure 37. Impaling the prey [TETZOMPAH, Loggerhead Shrike with prey on an agave] (Dibble & Anderson, 1963, Figure 145, after Paso y Troncoso).

"Fifth Paragraph," Miscellaneous Birds

Moving on to the "Fifth Paragraph, which telleth of still other kinds of birds, of whatever sort." It is unclear what logic of association might explain this list. Most are relatively small, though many stand out for bright colors, memorable vocalizations, or distinctive behaviors.

XOCHI-TŌTŌ-TL, literally, "flower bird," the Black-backed Oriole (*Icterus abellei*): "its throat, breast, belly are yellow: flower-like, well textured. It has a face-band. Its head, back, wings, tail are [black] mingled with white, in wavy line. Its legs are black" (FC, p. 45). Martín del Campo identified this as the "Bullock's Oriole," which has since been "split" taxonomically, with the resident Central Mexican form set apart as a distinct species, the Black-backed Oriole. The description is on point.

Ā-YACACH-TŌTŌ-TL, onomatopoetic, large wrens of the genus *Campylorhynchus*: "It is tawny. It is called *ayacachtototl* because its call, which it makes when it sings, is *cha cha cha cha, shi shi shi shi, charechi, charechi, cho cho cho cho*" (FC, p. 46). Martín del Campo identified this as the Band-backed Wren (*Campylorhynchus zonatus*). This particular species is at home on the Atlantic slope and might well have been known to the Aztec scribes. However, several other locally resident species of this same genus—equally noisy and conspicuous—might have been included without distinction: Gray-barred Wren (*C. megalopterus*), Rufous-naped Wren (*C. rufinucha*), Spotted Wren (*C. gularis*), Boucard's Wren (*C. jocosus*), and Cactus Wren (*C. brunneicapillus*). I would opt to equate this name with the genus rather than with any single species.

TACHITOHUIYA, onomatopoetic, perhaps the Green Shrike-Vireo (*Vireolanius pulchellus*) (Figure 38):

> It is small and green, small and round; a companion of the woodsman. It is named *tachitouya* because of its song, because its song says *tachitouya*. Whomever it sees, it comes along with him, singing as it goes; it goes along making [the sound] *tachitouya* (FC, p. 46).

There is no shortage of small, green woodland birds in Mexico, though few would fit the description. This bird's calls have been described as "a monotonously repeated, chanting *chew chew chew* or *chewy chewy chewy*" (Howell & Webb, 1995, p. 626) (cf. sound clip by Daniel Garrigues, ML 437064121). In my experience, vireos often seem to be following you, singing constantly in the treetops.

Figure 38. TACHITOHUYA, Green Shrike-Vireo (ML428104811, photo by William Hemstrom from Macaulay Library).

CUAUH-TOTOPO-TLI, literally, "tree piercer," the Golden-fronted Woodpecker (*Melanerpes aurifrons*):

> Also its name is *quauhchochopitli* <"tree pecker">, and its name is *quauhtatala* <"tree hitter">. The bill is pointed, pointed like a nail, strong, rugged, like obsidian. It is light ashen; agile; a tree-climber; a hopper up trees: a borer of holes in trees,…. Its food is worms; it destroys insects in trees when it bores them. There it nests (FC, p. 46).

Martín del Campo identified this as the Golden-fronted Woodpecker, a common and conspicuous species throughout Central Mexico. This may well be the prototypical woodpecker, but the name may have been applied as well to a number of other species of woodpeckers, such as the Golden-cheeked (*Melanerpes chrysogenys*), Gray-breasted (*M. hypopolius*), and Ladder-backed (*Dryobates scalaris*), all of which might have been described as "light ashen." Other common woodpeckers are predominantly black, brown, or olive, but otherwise fit the bill.

POXACUA-TL, literally, "stupid," the Mexican Whip-poor-will (*Antrostomus arizonae*): "It is like the Barn Owl; it looks like the barn owl. It is small, fluffy. As it flies, it only goes about flying erratically. Hence it is called *poxaquatl*. Dark yellow,… on the surface of its feathers" (FC, p. 46). This is most likely the Mexican Whip-poor-will. Other "nightjars" (Caprimulgidae) also fit the description but are less common and more restricted in range. These might have been included in the extended range of the category.

HUITLALO-TL, the Crested Guan (*Penelope purpurascens*): "It lives in the forest, like the wild turkey [*quauhtotolin*]. It is smoky, blackened. It is crested, but its crest is only of feathers" (FC, p. 46). The details would seem to nicely fit the Crested Guan.

CHĬCUA-TLI/TAPAL-CATZOTZON-QUI/CHĬCH-TLI, the Barn Owl (*Tyto alba*):

> It has thick feathers, eyes like spindle whorls, a curved bill. It is unkempt, fluffy. Its feathers are ashen, blotched like a quail's. It is round-headed, stubby tailed, round-winged. The eyes shine by night; they are weak by day. It is a night traveler which sees at night; it feeds, it lives by hunting. By day it is blinded by light…. It eats mice [and] lizards. It claws one…. It is named *tapalcatzotzonqui* because its call is as if one struck potsherd, or rattled them. Thus does it sound (FC, pp. 46–47).

This is certainly the Barn Owl, a common resident throughout Mexico. The rattle or "twitter" vocalization is given by males delivering food to the nest.

TLĀL-CHICUA-TLI, literally, "ground Barn Owl," the Burrowing Owl (*Athene cunicularia*):

> It is the same as the barn owl. It is small. For this reason is it called *tlalchi-quatli:* its nest is in a hole, underground, where it lays eggs sits, hatches it young. Only in a hole does it live; it is not a tree-climber, only a ground-dweller. It goes over the surface of the ground when it flies (FC, p. 47).

This description leaves no doubt that the Burrowing Owl is intended. The name is likely derived from TLĀLCHI "on or towards the ground" or TLĀL-LI "earth, land" (Kaarttunen, 1983, pp. 273, 275). Clearly, the Aztec scribes did not think the Burrowing Owl was "the same as" the Barn Owl, that phrase serving as shorthand for "similar." Note that the Common Gallinule (CUA-CHĪL-TON) and the American Coot (YACA-CIN-TLI) were also said to be "the same," despite contrasting descriptions. See ZACA-TECOLŌ-TL above, which may (or maybe not) be a synonym.

ILAMA-TŌTŌ-TL, literally, "old woman bird," the Canyon Towhee (*Melozone fuscus*): "It is ashen, chalky; ashen-backed. It is roundish, blunt-billed. It has a small crest. Its home is everywhere in settlements" (FC, p. 47). This non-descript, confiding sparrow-like bird is a common sight in towns throughout northern and central Mexico, replaced in Oaxaca by its relative, the White-throated Towhee (*M. albicollis*).

TLATHUICICI-TLI, onomatopoetic, the Canyon Wren (*Catherpes mexicanus*):

> It is the same as the <Canyon Towhee>. It is named *tlatuicicitli* because of its song. When it is still dark, long before dawn, it begins to sing. As it sounds its song, it is as if it says *tlatuicicitli*. It lives in one's roof, in one's wall. It awakens one (FC, p. 47).

Martín del Campo identified this as some species of wren of the genus *Thryothorus*. Four or five species of that genus might have been familiar to the Aztec scribes. However, in my experience, those wrens skulk in thick brushy tangles. By contrast, in my personal experience in Oaxaca, the Canyon Wren is conspicuous in villages and towns, foraging for insects in the adobe wall crevices and under eaves. Its characteristic descant song and scolds are un-mistakeable: "*see-see-see-syee-syee-syee-syee-syee-syee-syoo-syoo-syoo-syoo, jirr, jirr...*" (Howell & Webb, 1995, p. 561). This is a reasonable approximation of the sound-symbolic name (cf. sound clip by Brayden Luikart, ML436865601).

CHĪCUA-TŌTŌ-TL, literally, "Barn Owl bird," the Eastern Meadowlark (*Sturnella magna*): "The bill is pointed, the breast yellow. Its back, wings, tail are ashen, blotched like a quail, as well as its head" (FC, p. 47). I agree with Martín del Campo that this fits the Eastern Meadowlark best. When turned away so that the brilliant yellow breast and black chest band are invisible, it does resemble a small Barn Owl.

ZACA-TLA-TLI/ZACA-TLA-TON, literally, "[small] grass-eater," various sparrow species (Emberizidae): "The little sparrow is small and round, smoky. It is called *çacatlatli* because it lives in grasslands. Amaranth [seed] is its food" (FC, p. 47). As is often the case with folk biological classifications, small, non-descript birds may be "lumped" into broadly inclusive categories. This is clearly such a case, as literally dozens of species of such genera as *Peucaea*, *Aimophila*, *Amphispiza*, *Ammodramus*, *Melospiza*, and *Spizella* among others, might be included here. Some are residents, others are winter visitors.

TLAPAL-TŌTŌ-TL, literally, "red ink bird," Vermilion Flycatcher (*Pyrocephalus rubinus*): "Its body, its feathers are an over-all chili-red, but its wings, its tail are ashen,…. It is very chili-red, the color of dried chili. It is a night singer …. Four times, five times at night does it sing" (FC, p. 47). I agree with Martín del Campo's identification of this as the Vermilion Flycatcher. Though there are several bright red Mexican birds that approximate this description, this flycatcher is well known to sing at night, as noted (Howell & Webb, 1995, p. 505).

CHĪL-TOTOPIL, literally, "chili bird," possibly the Red Warbler (*Ergaticus ruber*): "It is the same as <that is, quite similar to> the [*tlapaltototl*] …. It has no blood; its blood is only like serous fluid" (FC, p. 48). Martín del Campo identified this as the Red Warbler. It is certainly "chili-red" but is not unique in that regard. The diminutive suffix points to its small size, which also fits the Red Warbler. This species seems the most likely candidate. It is common in the pine-oak forests of the central Mexican highlands.

MOLO-TL/CUA-CHICHIL/NŌCH-TŌTŌ-TL, the House Finch (*Haemorhous mexicanus*):

> It is chalky, ashen, dark ashen; short-billed; medium sized, small; agile, a hopper; a singer. It is a warbler, a talker. It is capable of domestication; it is teachable; it can be bred …. The completely ashen one is the hen, and the chili-red-headed one is the cock … because it is chili-red-headed it is named *quachichil* …. It is named *nochtototl* <literally, "*tuna* bird"> especially because its real food is tuna <fruit of the prickly-pear cactus>. It eats amaranth [seed], *chia*, ground maize … treated with lime (FC, p. 48).

CŌCO-TLI/CŌCOH-TLI, onomatopoetic, the Inca Dove (*Columbina inca*):

> It is small and squat, near the ground. The wings are spotted like *chia*, like quail, smooth. The legs are chili-red, short. And it is from its song that it is called *cocotli*; its song says, *coco, coco* …. It has only one mate. When [the mate] dies, it always goes about as if weeping, saying, *coco, coco*. And it is said that it destroys one's grief, that its flesh destroys one's torment and affliction. They make the jealous eat of its flesh; thus they will forget [their] jealousy (FC, p. 48).

This bird is still a comforting presence in every Mexican village and town.

"Sixth paragraph, which telleth of still other kinds of birds," but is, in fact, limited to the quail

ZŌL-IN/TECU-ZOLIN (the male)/OHUA-TON (the female), the Montezuma Quail (*Cyrtonix montezumae*):

> Its bill is pointed, ashen-green. Its breast is spotted with white; its wings are called *chia*-spotted. It is a runner …. When it lays eggs, when it hatches its young, it lays forty eggs. It hatches indeed fifty young … [*Tecuçolin*] is the cock. It is large, smoky-breasted, well spotted,… crested … [*Ooaton*] … is small, quite ashen, only a little spotted. This is the hen. Thus they nest: some make them wide; some cylindrical, so that they can sit on their eggs. And if they make them wide, she allows the companion to sit [with her] …. They can be bred, they can be domesticated (FC, p. 49).

There follow details of the quail's behavior; how the hen protects her young; how they are trapped in snares. According to Howell and Webb (1995, p. 229), Montezuma Quail lay 6–12 eggs. However, hens may adopt neighboring broods, which might have inspired the Aztec account. There are several additional species of quail that might have been familiar to the Aztec scribes: the Scaled Quail (*Callipepla squamata*), the Banded Quail (*Philortyx fasciatus*), and the Northern Bobwhite (*Colinus virginianus*). All are resident in the highlands of central Mexico. The Montezuma Quail favors somewhat higher elevations and the male is heavily spotted, so that species is likely the prototype. The extended range might include the other three species.

Eugene S. Hunn

"Seventh paragraph, which telleth of still other birds, of their habits"

TZANA-TL, the Slender-billed Grackle (*Quiscalus palustris*): "It is black. Its bill is well curved" (FC, p. 50). Though the description is brief, there is no doubt that this is the Slender-billed Grackle. It inhabited the highland marshes at the headwaters of the Lerma River near modern-day Toluca. These marshes have been largely drained for pasture, cultivation, and housing, leaving this species homeless. It was last reported alive here in 1910 and is presumed extinct (Howell & Webb, 1995, p. 766). It was a smaller version of the Great-tailed Grackle, treated below. There I raise the possibility that the Slender-billed Grackle could have descended from grackles intentionally established in the Valley of Mexico, a rather atypical habitat for these birds.

TEŌ-TZANA-TL, literally, "sacred grackle," the Great-tailed Grackle (*Quiscalus mexicanus*):

> It has a long, nail-like bill; it has a streaked tail…. It has a good voice; it speaks well, it speaks pleasantly. The one which is not very black, but a little sooty, is the hen; the very black one, very curved of bill, glistening, is the cockerel and is called *teotzanatl* …. It is named *teotzanatl* because it did not live here in Mexico in times of old. Later, in the time of the ruler Auitzotl it appeared here in <the Valley of> Mexico. For he commanded that it be brought here from [the provinces of] Cuextlan [and] Totonacapan <on the coast of Veracruz> (FC, p. 50).

There follows an account of how the species spread widely, protected by the emperor's dictate. Haemig (1978) provides a detailed discussion of this ancient avian transplantation. I find the Aztec appreciation of the vocal talents of this species somewhat at odds with my own reaction to this grackle's repertoire of "loud shrieks, clacks, whistles" (Howell & Webb, 1995, p. 740). Meso-american civilizations established great cities in the Valley of Mexico—notably at Teotihuacán—many centuries before the Aztecs founded their imperial headquarters a few miles south at Tenochtitlán. Murals preserved at Teotihuacán document the antiquity of the Mesoamerican fascination with "rich plumes" (Berrin, 1988). Thus, might the Slender-billed Grackle—like the Great-tailed, far from its usual lowland habitat—have represented an earlier introduction?

COYOL-TŌTŌ-TL, literally, "rattle bird"/ĀCA-TZANA-TL, literally, "reed grackle," Red-winged (*Agelaius phoeniceus*) and Yellow-headed Blackbirds (*Xanthocephalus xanthocephalus*), likely including also cowbirds (*Molothrus* spp.) in winter flocks:

> Some are quite black, some only smoky. They dwell among the reeds; among the reeds they hatch. They prey especially upon maize, and worms, and the small insects that fly … [*Coioltototl*] … is the same as the *acatzanatl*,

though some have a chili-red throat, breast, wings, rump <the Red-winged Blackbird>. Some are yellow-breasted, with white wing-bends <the Yellow-headed Blackbird>. And very good, very clear, is its song—much like a bell, pleasing, sweet <the Red-winged Blackbird>. Hence it is named *coyoltototl*. In the reeds, in the midst of the reeds it sits, it hatches (FC, p. 50).

HUĪLŌ-TL, onomatopoetic, the Mourning Dove (*Zenaida macroura*)

> It has a slender, pointed bill. It is chalky-ashen … the tail is long. It is some-what tall, long-necked. Its food is grains of maize, *chia*, amaranth, *argemo-ne, lepidium* seed …. It is lazy. Its nest is only sticks which it places together; it piles a little grass on the surface …. And its habits are much like the *cocotli* <the Inca Dove>. They are very attentive to their mates, to their hens. And if one dies, the one [remaining] always lives in mourning. And it seems constantly to weep; it makes [the sound] *uilo-o-o*. And its name, *uilotl*, is taken from its song, which says *uilo* (FC, p. 51).

There follows an account of a folk tale explaining the dove's derisive name, *tlamacazqui* "lazy." There can be no doubt that this is the Mourning Dove. The tale illustrates how people every-where find moral instruction watching birds.

TLĀCA-HUILO-TL, literally, "mister dove" <an approximation>, the Rock Pigeon (*Columba livia*): "It is large, round, ball-like. Some are ashen, some chalky, some dark green, some smoky, some tawny. It is really much like the Castilian *uilotl*" (FC, p. 51). Martín del Campo identified this bird as the Common Ground Dove (*Columbina passerina*). I fail to see the resemblance to the ground dove. Rather, given the diversity of plumages noted and the close comparison with the "Castilian *uilotl*," I am inclined to the view that this bird is, in fact, the introduced Rock Pigeon, that is, the "Castilian *uilotl*." it seems not unreasonable to conclude that feral Rock Pigeons in their characteristic variety of plumage colors might well have been established in Mexico City by 1560.

"Eighth paragraph, which telleth of the birds which are good singers"

CUĪTLACOCH-IN/CUĪTLACOCH-TŌTŌ-TL, onomatopoetic, the Curve-billed Thrasher (*Toxostoma curvirostre*):

> It has long legs, stick-like legs, very black; it has a pointed, slender, curved bill. It is ashen, ash-colored, dark ashen. It has a song, a varied song …. It is

named cuitlacochtototl, which is taken from its song, because it says *cuit-lacoch, cuitlacoch, tarata, tarat, tatatati, tatatari, titiriti, tiriti…* <cf., sound clip by Andrew Theus, ML436112051>. It is capable of domestication; it is teachable. It breeds everywhere, in treetops, in openings in walls. Wherever it is inaccessible, there it breeds. Its food is insects, flies, water flies, flesh, ground maize (FC, p. 51).

There is no doubt as to the identity of this species. It is still known in rural Mexico as *cuicacoche* (Schoenhals, 1988, p. 388). The culinary delicacy—the parasitic fungus (*Ustilago maydis*) that infests maize ears—is also known as "*huitlacoche/cuitlacoche*," "raven's excrement." This is apparently a near coincidence, as the bird's name is derivative of *cuīcatl* "song," rather than *cuicatla* "excrement." "Song" has a long "ī," "excrement" a short "i" (https://en.wikipedia.org/wiki/Corn_smut). One must be careful to enunciate in Náhuatl.

ZENTZON-TLAHTŌLEH, literally, "four-hundred speaker," the Northern Mockingbird (*Mimus polyglottos*):

It is ashen, a little dusky. The breast is white, the wings are white mingled with red. There is a [white] bar over the eyes. [The body] is small and long. Its dwelling place, where it breeds, is in the forest, in inaccessible places. It does not sing in the winter, only during the summer…. It is named *centzontlatole* because it mocks all the birds; it also mocks the turkeys – the cocks, the hens. It also mocks the dogs (FC, p. 52).

The Northern Mockingbird fits the bill, though some have suggested this might refer to the Brown-backed Solitaire (*Myadestes occidentalis*), another famous avian diva. However, the plumage details (with the exception of the "red" in the wing) and the mockery clearly point rather to the mockingbird.

MIYAHUA-TŌTŌ-TL, literally, "maize tassel bird," the Lesser Goldfinch (*Spinus psaltria*): "Also its name is *xopan tototl* <literally, "summer bird">. It is small, round, dark yellow. It has a song, a soft song; it is a singer. It gladdens one; it makes one rejoice. It is small, tiny, minute" (FC, p. 52). I believe this is the Lesser Goldfinch (*Spinus psaltria*) based largely on my experience with this species in a Zapotec-speaking town in Oaxaca, where the Lesser Goldfinch was likewise known as "maize-tassel bird" (Hunn, 2008). It is a common resident throughout Mexico and matches the details.

CHIQUIMO-LI/CHIQUIMO-LIN, most likely the Imperial Woodpecker (*Campephilus imperialis*) (Figure 39):

> It is as large as a *tzanatl*. It is crested. Dark colored is its crest; white its bill; black spotted with ashen are its feathers. Its throat is yellow…. Its food is tree worms; it extracts the worms from the trees. And it nests, it breeds within the tree; it makes a hole in the tree…. And when it sings, it cries out much, it warbles, sometimes like the whistling with the fingers; and it sings as if there were many birds…. But when it seems to shriek, it is angry. So it is said, this was taken as an omen…. "take care, something may befall us." But when it whistled, they said it was happy…. And when there is contention, one is called *chiquimolin* for this reason (FC, p. 52).

Martín del Campo identified this as the Ladder-backed Woodpecker (*Dryobates scalaris*), which is one of the most common and widespread woodpecker species in central Mexico. However, that species is small and has no crest, its bill is dark, nor does it vocalize as described. Clearly this is a species of woodpecker, but which one? Are there crested woodpeckers with white bills? The Pale-billed Woodpecker (*Campephilus guatemalensis*) is widespread along both coasts but absent from the highlands. One other possibility comes to mind, the Imperial Woodpecker (*Campephilus imperialis*), once the largest woodpecker in the world. It is now considered extinct due to habitat destruction, as it was last reported (and filmed!) in 1956 (Lammertink et al., 2011). However, it is known to have ranged widely in historic times throughout the pine-oak forest of the Sierra Madre Occidental, from Chihuahua state south to Jalisco and Michoacan, nearly to the Valley of Mexico. It would seem a bird the Aztecs would not have overlooked. Its calls have been recorded as "nasal penny-trumpet-like notes" (Howell & Webb, 1995, p. 766), though that may be just a sample of an extensive repertoire.

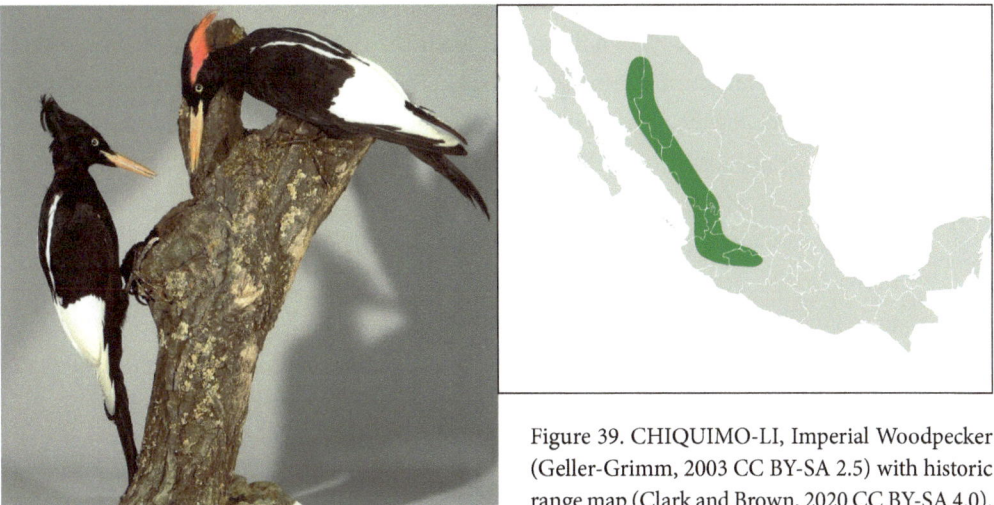

Figure 39. CHIQUIMO-LI, Imperial Woodpecker (Geller-Grimm, 2003 CC BY-SA 2.5) with historic range map (Clark and Brown, 2020 CC BY-SA 4.0).

CHACHALACAME-TL, onomatopoetic, prototypically, the West Mexican Chachalaca (*Ortalis poliocephala*):

> It is the same size as the *teotzanatl* <Great-tailed Grackle>. It is somewhat yellow all over; its tail is mixed [black and] white. Its food is fruit; also maize kernels, ground maize. It nests in inaccessible parts of trees. And it sings in winter. It is called *chachalacametl* because if a number of them settle together, only one begins to sing; then all sing. And the neck is like a turkey's neck, only very small .... And it sings three times during the night .... It is said that it awakens one (FC, p. 53).

Martín del Campo identifies this as the Plain (Common) Chachalaca (*Ortalis vetula*). It is certainly a species of chachalaca, known throughout Mexico today by that onomatopoetic name. However, there are three species that the Aztecs might have encountered. The Plain Chachalaca is a bird of the Atlantic slope lowlands. Rather more likely to have been well known to the Aztec scribes is the West Mexican Chachalaca (*Ortalis poliocephala*). No doubt, all species would have been so-named in their respective home territories.

"Ninth paragraph, which telleth of the native turkeys"

TŌTO-LI/TŌTO-LIN, the domestic Wild Turkey (*Meleagris gallopavo*) (Figure 40):

> Its name is also *iuiquen, iuiquentzin* <literally, "the one with the feathered mantle">, and *xiuhcozca* <literally, "the one with the turquoise collar"> is [also] its name. It is a dweller in one's home, which can be raised in one's home, which lives near and by one. The feathers are thick, the tail rounded .... it is heavy, not a flyer .... It leads the meats; it is the master .... Some turkeys are smoky, some quite black,... some white, some ashen, ash-colored, some tawny, some smoky ... <plumage variation indicates domestication>. The cockerel, the male turkey <"HUEHXŌLŌ-TL," literally, "old servant">, is big, big and coarse. It has air-sacs, a big belly, wattles,... a long neck. The stalky neck has a necklace, a neck-coral. The head is blue;... it has a dewlap. The eyelids are cylindrical, swollen. It has a rounded protuberance, an erectile process. It is a tail-feather-spreader, a bristler, an attacker. It ... sounds its air-sacs .... Its rounded protuberance is pliant, leathery,... soft, very soft. One who hates another feeds <the protuberance> to him in chocolate, in sauce; he causes him to swallow it. It is said that he thereby makes one impotent (FC, pp. 53–54) (Figure 40).

Figure 40. TŌTO-LIN, turkeys (Dibble & Anderson, 1963, Figure 170, after Paso y Troncoso).

CUAUH-TŌTO-LIN, literally, "forest turkey," the wild Wild Turkey (*Melagris gallopavo*): "It is like the domesticated turkey, though a little smoke-colored. The wing-bend is white…. It is a gobbler. The turkey cock [and] the turkey hen … they gobble, they cough …." (FC, p. 29). This species was listed with the waterfowl due, presumably, to the fact that the American White Pelican was known as the "water turkey" (see above). Wild Turkeys continue to survive in northern Mexican highland forests, though clearly distinguished from their captive brethren. Martín del Campo suggested that this term might apply to the Ocellated Turkey (*Meleagris ocellata*) (1940, p. 395). That species is well known to the Maya, but is unknown north of the Yucatan peninsula, thus not likely recognized by the Aztecs.

"Tenth paragraph, which telleth of the parts of the different birds"

The Aztec knew the anatomy of a bird in fine detail, particularly, the feathers tracts, central as this knowledge was to the feather workers, the *amanteca*. Sixteen named types of body feathers are listed, then six wing-feather tracts are distinguished. The parts of a feather are named. Then the wing, tail, bill, eye, eyelid, head, brain, skull, neck, tongue, throat, gullet, belly, intestines, crop and gizzard are each named, with variations noted. Finally, the ovary, rump gland, and toe are also named.

Having feathers was considered the defining characteristic of a "bird": "The property, the possession, which belongs to all the different birds and to turkeys is feathers <IHHUI-TL>" (FC, p. 54). There is a suggestion here that turkeys are, at least, a most extraordinary bird, as shown by the extensive details described in Paragraph Nine.

Eugene S. Hunn

Many classes of feathers are named, depending on their respective feather tracts or their notable ornamental values:

> Those <feathers> which appear on the head of a resplendent quetzal are called *quetzaltzinitzcan*. And those which appear on the neck are called *tapalcayotl*; its *tlapalcayotl* feathers…." Those which appear on its belly and on its back are called *alapachtli*…. Those which are right on its skin <"downy feathers"> are called *tlachcayotl*…. So one refers to the *tlachcayotl* feathers of the eagle, the scarlet macaw, the *xomotl* <House Finch>. And those which are at the edge of its rump, which cover the base of the tail <"upper tail coverts"> are called *olincayotl, poyaualli, poyauallotl* <cf. POYĀHU(A)- "to darken" (Karttunen, 1983, p. 204)> (FC, p. 54–55).

ZA-CUAN, QUECHŌL, and TZINITZCAN name particular bird species but also name the distinctive plumes of these "precious ones": brilliant yellow of the Montezuma Oropendola, passionate pink of the Roseate Spoonbill, and glistening green of the Mountain Trogon. TZINITZCAN also indicates head plumes: "And those which appear on their heads, even the not precious, are called *tzinitzcan*" (FC, p. 54). It seems that different types of feathers may be distinguished on the basis of either their location or their qualities or some combination of these features.

Discussion

Cultural Values of Birds

Birds were clearly objects of fascination and admiration for the Aztec people. Their aesthetic appreciation of the birds' brilliant plumage and of their varied vocalizations comes across loud and clear in the *Codex* testimony. Birds were highly valued sources of "precious feathers," woven into dramatic costume displays reserved for the high nobility as symbols of their near-god-like powers. The emperor Moteuczoma Xocoyotzin, whose fateful encounter with Hernán Cortés presaged the destruction of the Aztec civilization, is shown crowned by a headdress of radiating emerald-green Quetzal tail streamers framed by bands of tightly woven feathers of sky-blue from the Lovely Cotinga, passionate pink from the Roseate Spoonbill, and milk-chocolate from the Squirrel Cuckoo (see Figure 6, from the Kunsthistorisches Museum, Wien). These exotic plumes came in tribute from subject provinces to the east and south, as far as present-day Guatemala. Tribute lists record quantities of 400 or 8,000 such feathers (they shared the Mesoamerican vigesimal counting scheme) sent to the Aztec capital (Berdan & Anawalt 1992), to be transformed into elaborate costumes by professional feather-weavers of the the *amanteca* (ĀMANTĒCA-TL) guild.

Birds Valued for their Feathers (13, plus 7 more implied):
Birds of rich (or "precious") plumage ("*aves de pluma rica*"), listed in the first paragraph:

- QUETZAL-TŌTŌ-TL, Resplendent Quetzal;
- TZINITZCAN-TŌTŌ-TL, Mountain Trogon;
- TLĀUH-QUECHŌL, Roseate Spoonbill;
- XIUH-QUECHŌL, Lesson's Motmot;
- XIHUA-PAL-QUECHŌL, Turquoise-browed Motmot;
- ZA-CUAN, Montezuma Oropendola [= "troupial"];
- AYO-CUAN$_1$, Yellow-winged Cacique;
- AYO-CUAN$_2$, Northern Jacana;

- CHALCHIUH-TŌTŌ-TL, Red-legged Honeycreeper;
- XIUH-TŌTŌ-TL, Lovely Cotinga;
- XOCHI-TENACAL, Keel-billed Toucan;
- CUAPPACH-TŌTŌ-TL, Squirrel Cuckoo;
- ĒLŌ-TŌTŌ-TL, Blue Grosbeak.

Parrots and their relatives and hummingbirds were clearly appreciated for their colorful plumage though they were not specified as sources of decorative plumage in Book 11 of the *Codex*. However, these and other birds were noted as sources of precious feathers elsewhere in the *Codex*. The most detailed references appear in Chapter 12 of Book Eight of the *Codex*, which details "the decorations that the lords used in war." Here it is clear that "Nahuas associated precious feathers with rulers and warriors," the Aztec scribes describing the feathers from "the red spoonbill, the quetzal, the blue cotinga, the parrot, the troupial, and the heron that the rulers wore when they went to war" (Terracino, 2019, p. 49). Diego Durán described a ritual food prepared for a major feast:

> After shaping the dough [of corn-amaranth] into a *teixiptla*, or image of the deity in the form of a person, they place feather down and a hummingbird totem on the head, a feather arrangement called *anecuiutl*, a ball of parrot feathers at the back of his neck, a cloak feathered with eagle down, and a shield feathered with eagle down (Terraciano, 2019, p. 54).

Parrots were valued also for their vocalizations. So, we may add the following birds to the list of those with "precious plumage."

- TOZ-NENE, young Yellow-headed Amazon;
- TOZ-TLI, adult Yellow-headed Amazon;
- ALO, Scarlet Macaw;
- COCHO, White-fronted Amazon;
- QUILITON, Orange-throated Parakeet;
- TLALA-CUEZA-LI, Red-crowned Amazon;
- HUĪTZITZIL-IN, hummingbirds in general;
- QUETZAL-HUĪTZITZIL-IN, Garnet-throated Hummingbird, color scheme reminiscent of the Quetzal;
- XI-HUĪTZITZIL-IN, an unidentified hummingbird species noted for its turquoise plumage;
- TLE-HUĪTZIL-IN, another unidentified hummingbird with "glistening, resplendent" feathers;

A few additional species likely provided ornamental plumage for the Amanteca artisans:

- AZTA-TL, Snowy Egret, also known as TEO-AZTA-TL, sacred egret, with "intensely white" plumage;
- CUĀUH-TLI, eagles and large hawks in general: "the different eagles ... <were hunted; the young captured from nests and raised by hand>"
- ITZ-CUĀUH-TLI, Golden Eagle: "<feathers called *quauhxilotl*> glisten as if blotched with gold ..."

Birds for Food

The Aztec capital, Tenochtitlán, was established by the impoverished immigrant Aztecs in the 14th century on an uninhabited marshy island in Lake Texcoco (see Figure 1). They reclaimed the muddy margins by constructing *chinampas*, so-called "floating gardens," surrounding the city core of temples and elite residences. They proceeded to subjugate neighboring city states, and eventually, as a "triple alliance," extended their control by force of arms to encompass much of contemporary Mexico. The Aztec scribes employed by Sahagún for his encyclopedic project, were likely born and raised in or very near what remained of Tenochtitlán, this "Venice of America," in the decades following 1521.

The Aztec capital was actually a twin city. Tenochtitlán was the southern division, the focal point of administrative and ritual action, oriented to the paired great temples dedicated to Tetzcatlipoca and Huitziliopochtli. Here war captives were sacrificed to feed blood to the Sun in great ceremonies. The palaces of the high nobility surrounded this central ritual precinct. The northern division, Tlatelolco, was the site of a great marketplace where produce of all sorts arrived by canoe through the complex of canals that linked the central market with the multitude of cities and towns on the shores of the five lakes filling the Valley of Mexico. The Aztec capital is judged to have housed 200,000 to 300,000 citizens (Coe, 1962, p. 161), on par with the greatest European cities of the time. Clearly, to support such a population—including administrators, priests, soldiers, artisans, and assorted workers—required importing food in quantity.

The prominence of "waterfowl" in the *Codex* inventory of birds, points to the key role played by "water folk" as consultants in this project. These "water folk" (Ā-TLĀCA-TL, literally, "person made of water" [cf. Karttunen, 1983, p. 13]), were apparently peasant communities specializing in harvesting food and other resources from the extensive Valley of Mexico marshes for export to the capital. They hunted with spears, nets, and snares from canoes (Figures 41 and 42), harvesting not only birds, but also fishes and the giant larval salamander, the *axolotl* (Ā-XŌLŌ-TL). These waterfowl, these "animals which dwell in the water," are described in detail in the third chapter of Book 11 of the *Codex*. The majority of the waterfowl hunted for food were winter visitors, present in huge numbers from fall through spring. Summer harvests apparently shifted to fish, which abounded with the

Figure 41. Water Folk hunting the Ā-CIH-TLI, Clark's/Western Grebe (Dibble & Anderson, 1963, Figure 87).

Figure 42. Netting birds (Dibble & Anderson, 1963, Figure 187).

rains. Several resident species of waterfowl are noted as omens forecasting heavy rains, and thus an abundance of fish.

Some 30 birds were said to be edible, including 13 ducks, two grebes, a bittern, a heron, an ibis, a crane, two rails, a gull, and a pelican. The water folk feared the pelican. They sought to capture pelicans in hopes of securing a precious jade stone from its gizzard. However, in attempting to capture the bird, it could overturn their canoes and drown them (Figure 43). They also feared the cormorant for the same reason.

Figure 43. Capturing Ā-TŌTO-LIN, American White Pelican (Dibble & Anderson, 1963, Figure 84).

Most birds noted as edible were waterfowl. Exceptions also noted as edible were two tinamous, wild and domestic turkeys, quail, and the Inca Dove, the last named consumed only as a medicine. The Peregrine Falcon and the Red Warbler were explicitly noted as inedible.

Thirty species of birds are specifically noted as edible, including two also used for ritual/medicine; plus four possibly edible, and five explicitly inedible:

- YOLLO-TŌTŌ-TL, Rose-throated Becard, "edible" ["*qualoni*" <CUĀ-LŌ-NI>]; "when we die, our hearts turn into [these birds]";
- PŌPOCALES, Russet-naped Wood-Rail, "edible";
- TECUZIL-TŌTŌ-TL, Thicket Tinamou, "edible";
- IXMATLA-TŌTŌ-TL, Great Tinamou, "edible";
- TŌTŌ-MEH, "*muchipa atla nemj* <water bird, in general>, all are hunted, caught…. netted, noosed,… snared";
- TLALALACA-TL, Black-bellied Whistling-Duck, "good-tasting, savory" ["*velic, aviac*"];
- TOCUIL-COYŌ-TL, Sandhill Crane, "edible, savory, of good taste" ["*veltic*"];
- Ā-TZITZICUILO-TL, Black-necked Stilt, "is heavily fleshed, fat, greasy" ["*nacateton, chiaoac, vel ceceio*,"];
- CUAUH-TŌTŌ-LIN, Wild turkey, "edible – savory, good-tasting, fat";
- Ā-TŌTŌ-LIN, American White Pelican, ritual food;

- Ā-COYO-TL, Neotropic Cormorant, said to be the same as A-TŌTO-LIN, the pelican;
- Ā-CIH-TLI, Clark's or Western Grebe, caught in nets; speared; summons the wind;
- COHUIX-IN, Aztec Rail, "its flesh is edible" ["*qualonj in jnacaio*"];
- ICXI-XOXOUHQUI, American Avocet, "It is also edible" = "no qualonj";
- QUETZAL-TEZOLOC-TON, Green-winged Teal, "Its flesh is edible";
- METZ-CANAUH-TLI, Blue-winged Teal, "Its flesh is edible";
- CUA-COZ-TLI , Canvasback, "Its flesh is very savory" ["*cenca velic in jnacaio*"];
- ECA-TŌTŌ-TL, Wood Duck, "It is edible";
- AMANACOCHE, Bufflehead, "They are edible";
- Ā-TAPALCA-TL, Ruddy Duck, "They are edible";
- TZITZIHUA, Northern Pintail, "Their flesh is edible, good-tasting, savory, [and] not fishy" ["*amo xoqujiac*"];
- XĀL-CUĀ-NI, American Wigeon, "They are edible, savory";
- YACA-PITZA-HUAC, Eared Grebe, "It is not fishy, its flesh is savory";
- TZON-YAYAUHQUI, Lesser Scaup, "Good tasting is their flesh; it is fat, like bacon";
- ZŌL-CANAUH-TLI, Gadwall, "Good tasting is their flesh";
- CHIL-CANAUH-TLI, Cinnamon Teal, "They are edible";
- Ā-CHALALAC-TLI, Belted Kingfisher, "They are edible";
- YACA-PATLA-HUAC, Northern Shoveler, "It is edible";
- HUĀC-TLI₁, Black-crowned Night-Heron, "It is edible, savory";
- PIPITZ-TLI, Laughing Gull, "It is edible";
- ĀCA-CHICHIC-TLI, Least Bittern, "It is edible";
- ZŌL-IN, Montezuma Quail, "edible, savory ... exceedingly good tasting" ["*vel tzompa-lalatic*"]; domesticated;
- TŌTO-LIN, Wild Turkey, domesticated, "it is edible; it leads the meats .... It is tasty, fat, savory."
- TLAPAL-TŌTŌ-TL, Vermilion Flycatcher, "It is not fat" ["*amo chiaoa*"], but neverthe-less "es bueno de comer" (García Quintana & López Austin, 1988, p. 709);
- CŌCOH-TLI, Inca Dove, ritual food: "its flesh destroys one's torment and afflic-tion ... they will forget their jealousy."

Birds specified as inedible (some only in the *Historia General*) include the following:
- TZOPĪLŌ-TL, vultures, "No son de comer" (García Quintana & López Austin, 1988, p. 706);
- NECUILIC-TLI, Merlin, "No es de comer" (García Quintana & López Austin, 1988, p. 707);
- CHĪL-TOTOPIL, Red Warbler, "Its flesh is inedible .... It has no blood";
- TZANA-TL, Slender-billed Grackle, "no son buenas de comer" (García Quintana & López Austin, 1988, p. 710);

- ĀCA-TZANA-TL, various blackbirds and cowbirds, "no son de comer" (García Quintana & López Austin, 1988, p. 711).

Bird Voices

The Aztecs lived in what might be characterized as a more-than-human society, that is, they were open to meaningful communication with birds and other "earthly things." They attributed power to speech. The Aztec emperor was the Tlatoani (TLĀTOA-NI), "speaker, one who commands." So it is perhaps not surprising that they enjoyed raising parrots and parakeets and other "birds which are good singers." These are listed in the second and eighth paragraphs of the bird chapter of Book 11 of the *Codex* (FC, pp. 22 and 51). They said of the White-fronted Amazon that, "It is a singer, a constant singer, a talker, a speaker, a mimic, an answerer, an imitator, a word-repeater. It repeats one's words, imitates one, sings, constantly sings, chatters, talks" (FC, p. 23). The Aztecs also hand-raised House Finches (MOLO-TL): "it is a warbler, a talker. It is capable of domestication; it is teachable" (FC, p. 48).

The emperor Ahuitzotl, predecessor to Moteuczoma Xocoyotzin, ordered that Great-tailed Grackles be brought from the Veracruz coast and released in Tenochtitlán, where they were to be fed and protected. Ahuitzotl (1440–1502 AD) admired the bird for "its good voice; it speaks well, it speaks pleasantly." To my ear, the grackle hardly "speaks pleasantly," but tastes vary.

Paragraph Eight of Chapter 2 of Book 11 is devoted to the "birds which are good singers" ["*itechpa tlatoa in tototme, vel cujcanj*"]. These included the Curve-billed Thrasher and the Northern Mockingbird. The thrasher (CUĬTLACOCH-IN) "has a varied song…. It is capable of domestication; it is teachable…. And in winter it does not sing…. When the rains come,… when it becomes warm, then it begins to sing" (FC, pp. 51–52). Its name mimics its song. The mockingbird likewise "does not sing in the winter, only during the summer…. it mocks all the birds… the turkeys…. the dogs…. it sings all night" (FC, p. 52). The association of bird song with the rains and summer is also noted for the Lesser Goldfinch, MIYAHUA-TŌTŌ-TL, literally, "bird of the maize tassel." It is also known as XOPAN-TŌTŌ-TL, "summer bird." The last two "good singers" are the Imperial Woodpecker, CHIQUIMO-LIN, and the West Mexican Chachalaca, CHACHALACAME-TL. They said of the chachalaca, "It is called *chachalacametl* because if a number of them settle together, only one begins to sing; then all sing…. it sings three times during the night…. It is said, it awakens one" (FC, p. 53).

The Imperial Woodpecker, now very likely extinct, was once the largest woodpecker in the world. Like its close relative, the Ivory-billed Woodpecker, it was not known for vocal power. On the contrary, its calls were described as like a "nasal penny-trumpet" (Howell & Webb, 1995, p. 766). The Aztecs described it rather differently:

> when it sings, it cries out much, it warbles, sometimes like whistling with
> the fingers; and it sings as if there were many birds…. But when it seems

to shriek, it is angry. So, it is said, this is an omen.... And when there is contention, one is called *chiquimolin* for this reason.... When one arouses contentions, when one arouses bad feelings among others (FC, pp. 52–53).

The "good singers" and other voices (18 species):

- YOLLO-TŌTŌ-TL, Rose-throated Becard, "when we die, our hearts turn into [these birds]"; "when it sings...it gladdens";
- PŌPOCALES, Russet-naped Wood-Rail, "Always at twilight and at dawn, it says *popocales*";
- IXMATLA-TŌTŌ-TL, Great Tinamou, "When it sings... <it is> as if it imitated those who live there";
- CUĪCUĪTZCA-TL, Barn Swallow, "It awakens sleepers";
- TOL-COMOC-TLI, American Bittern, sounds like a "two-toned drum"; forecasts rain;
- HUAC-TLI$_2$, Laughing Falon, "its laugh sounds like a man";
- TACHITOHUIYA, Green Shrike-Vireo, "a companion of the woodsman" ["*quauhtla-catl, tevivicanj*"];
- TLATHUICICI-TLI, Canyon Wren, "It awakens one,"
- TLAPAL-TŌTŌ-TL, Vermilion Flycatcher, sings at night;
- MOLO-TL, House Finch, "It sings...constantly"; captured and domesticated for its song;
- TEŌ-TZANA-TL, Great-tailed Grackle, "it speaks well"; introduced to the Valley of Mexico by Ahuitzotl and fostered;
- COYOL-TŌTŌ-TL, Red-winged and Yellow-headed Blackbirds, "very good, very clear is its song—much like a bell, pleasing, sweet";
- HUĪLŌ-TL, Mourning Dove, "It is lazy"; very attentive to their mates; song is like "weeping" for lost mate;
- CUĪTLACOCH-IN, Curve-billed Thrasher, "it has...a varied song"; "good singer" ["*vel cujcani*"]; domesticated; sings with the rains;
- ZENTZON-TLAHTŌLEH, Northern Mockingbird, "a good singer"; mocks all the birds...also...the dogs";
- MIYAHUA-TŌTŌ-TL, Lesser Goldfinch, "good singer"; "It has...a soft song.... It gladdens one; it makes one rejoice";
- CHIQUIMO-LIN, Imperial Woodpecker, "good singer"; "when it seems to shriek, it is angry"; an omen;
- CHACHALACAME-TL, West Mexican Chachalaca, "good singer"; "it awakens one."

Birds as Omens, Medicines, and Cultural Icons

For the Aztecs, birds might have a lot to say. They indicate the passage of the seasons, the onset of rain or of frost. They may foretell good or bad fortune, for a person or for the rulers. Consider the Rose-throated Becard, YOLLO-TŌTŌ-TL, literally, "heart bird."

> It is quite small.... As for its being called *yollototl* <"heart bird">, the people there say thus: that when we die, our hearts turn into [these birds]. And when it speaks, when it sings, it makes its voice pleading; it indeed gladdens one's heart, it consoles one (FC, p. 25).

By contrast, the Wood Stork, CUA-PETLĀ-HUAC, is considered an evil omen, for "when it is caught,... Either some of the lords will die, or war [will come] .... Thus did the water folk verify it: as often as they took wood <stork>, so often the city suffered some harm,..." (FC, p. 32). The mysterious bird known as TEN-ITZ-TLI, "flint beak," is likewise feared:

> if one captures this <flint-beak>, it is a sign that he is about to die. And they say that his house will be destroyed; his house folk will perish. For this reason, it is named a bird of ill omen <TĒTZĀHUIĀ-TŌTŌ-TL> (FC, p. 31).

The description seems bizarre—nocturnal, flying high over the lagoon hunting water flies, with three bills and two tongues—but I suspect this could be the Black Tern, which hawks insects over the marshes, often at dusk. Since they feared to capture it, they likely never got a close look, which may help account for the strange features of their description.

The American Bittern

> is always a portent <for the water folk>. When it sings a good deal <booming like a log-drum>, always all night, they know thereby that rains will come, it will rain much, and there will be many fish—all manner of water life. But if it will not rain much,... it does not sing much. Only perhaps every third day ... does it sing (FC, p. 33).

Its little brother, the Least Bittern, ĀCA-CHICHIC-TLI, "is also an omen for the water folk: always when it sings it is about to dawn. First it begins to cry out; then the various [other] waterfowl answer it" (FC, p. 39). Another dawn singer is the Canyon Wren, TLATHUICICI-TLI. "When it is still dark, long before dawn, it begins to sing. As it sounds its song, it is as if it says *tlatuicicitli*, <which the locals interpreted as, "Hello! Hello! Already it is dawn" (García Quintana & López Austin, 1988, p. 709)>. It lives in one's rafters, in one's wall. It awakens one" (FC, p. 47).

Finally, a few birds were medicinal. For example, the Inca Dove, COCOH-TLI:

> It has only one mate. When [its mate] dies, it always goes about as if weeping, saying, *coco, coco*. And it is said that it destroys one's grief, that its flesh destroys one's torment and affliction. They make the jealous eat of this flesh; thus they will forget [their] jealousy (FC, p. 48).

The hummingbird, HUĪTZITZIL-IN, "is medicine for pustules. One who wishes never to have pustules eats often of its flesh. But they say it makes one sterile" (FC, p. 24). An early example of anti-vax sentiment? Finally, the turkey is prescribed: "Its rounded protruberance is pliant, leathery,… soft…. One who hates another feeds <the protruberance> to him in chocolate, in sauce; he causes him to swallow it. It is said that he thereby makes one impotent" (FC, pp. 53–54).

Birds as medicine, omens, or cultural icons (15 species plus three that are additional possibilities):

- HUĪTZITZIL-IN, hummingbird, "is medicine for pustules";
- YOLLO-TŌTŌ-TL, Rose-throated Becard, "when we die, our hearts turn into [these birds]"; "when it sings… it gladdens";
- Ā-TŌTO-LIN, American White Pelican, ritual food; captured in hopes of finding precious stones in the gizzard; calls wind; sinks boats;
- Ā-COYO-TL, Neotropic Cormorant, said also to "sink one" and is otherwise like the Ā-TŌTO-LIN;
- Ā-CIH-TLI, Clark's or Western Grebe, caught in nets; speared; summons the wind;
- TEN-ITZ-TLI, Black Tern, "bird of ill omen" ["*tetzauhtototl*"]; if one is captured it is a sign the person will die;
- CUA-PETLA-HUAC, Wood Stork, "when it is caught, it is considered an evil omen" cf. "*tetzahuitl*," that is, "something extraordinary, frightening…" (Karttunen, 1983, p. 237);
- CUA-TEZCA-TL, Purple Gallinule, literally, "mirror-head"; "mirror" on its head is where one's future is shown; a "sign of war"; if captured;
- TOL-COMOC-TLI, American Bittern, sounds like a "two-toned drum"; forecasts rain;
- ĀCA-CHICHIC-TLI, Least Bittern, "an omen for the water folk: always when it sings it is about to dawn";
- CĀCĀLŌ-TL, Common Raven, perhaps just a crop raiding pest;
- CUĀUH-TLOH-TLI, Peregrine Falcon, illustrated being captured; "gives life to Uitzilopochtli" <the patron god of the Aztecs> with blood;
- NECUILIC-TLI, Merlin, "from the wind it knows when the frost is about to come. Then it begins to sing";
- CŌCOH-TLI, Inca Dove, ritual food: "its flesh destroys one's torment and affliction… jealousy";

- HUĪLŌ-TL, Mourning Dove, "It is lazy"; very attentive to their mates; song is like "weeping" for lost mate;
- CUĪTLACOCH-IN, Curve-billed Thrasher, "it has…a varied song"; domesticated; "sings with the rains";
- CHIQUIMO-LIN, Imperial Woodpecker, "when it seems to shriek, it is angry"; an omen;
- TŌTO-LIN, Wild Turkey, domesticated, "it is edible; it leads the meats"

The Aztec Ornithological Taxonomy

One central focus of ethnobiological studies involves comparing the taxonomic structure and its nomenclatural correlates across cultures. Brent Berlin (1992) has developed a comprehensive and systematic framework for such comparisons. Both ethnobotanical and ethnozoological "kingdoms" exhibit familiar structural and nomenclatural features, and the Aztec Náhuatl ornithological classification is no exception. Berlin's framework posits a set of universal ethnobiological taxonomic ranks. These are, in descending order of inclusiveness: Kingdom (e.g., "plant," "animal"), Lifeform (e.g., "tree," "bird"), Intermediate (e.g., "fruit tree," "duck"), folk generic (e.g., "oak," "wren"), folk specific (e.g., "live oak," "house wren"), and folk varietal (e.g., "coast live oak," "southern house wren"). Intermediate rank taxa are uncommon and may be absent and are often of a "hybrid" character, such as when a utilitarian consideration groups together generics of quite distinct appearance. Varietals are also rare and typically involve domesticated species and/or those of high cultural salience.

As a general rule, the great majority of folk taxa (that is, categories of organisms "that go together") are what Berlin has called "folk generics." Berlin's terminology here has generated some confusion, in that the categories he labels folk generic more often than not correspond to individual species in the Linnaean scheme (cf. Hunn, 1977, p. 61, Table 3.5). He argues that such taxa tend strongly to have relatively simple names, only occasionally of "binomial form," unlike the names of Linnaean genera. Binomial (and occasionally trinomial) names most often name "folk specific" taxa, which are "kinds of" a folk generic. Binomial names include two parts, a "head" term that labels a more inclusive category modified by an attributive (e.g., as in English "white pine" or "bald eagle," the first a kind of "pine," contrasting with "lodgepole pine," the second a kind of "eagle," contrasting with "golden eagle"). Though binomial names are particularly characteristic of folk specific taxa where the head term names the superordinate folk genus, they may also name folk genera when the head term names an inclusive lifeform or an intermediate category. Examples from English include "apple tree" and "hummingbird," for which "tree" and "bird" name lifeforms.

The Aztec folk taxonomy of birds as reported in the *Florentine Codex* fits comfortably within Berlin's framework, as we shall see. The Aztec understanding of how birds fit within a larger scheme may be clearly seen in "Book 11—Earthly Things" of the *Florentine Codex.* This book… "telleth of the different animals, the birds, the fishes; and the trees and the herbs; the metals resting in the earth—tin, lead, and still others; and the different stones" (FC, p. 1). The first chapter "telleth of the four-footed animals" ("iolque" = YŌLQUI, all of which are mammals); the second chapter "telleth of all the different kinds of birds" ("totome," plural of TŌTŌ-TL); the third chapter "telleth of all the animals which dwell in the water," including in the second paragraph, "all the fishes" ("mjmjchtin"); the fourth chapter "telleth of of still other animals which live in the water, which are inedible"; the fifth chapter "telleth of the various serpents,…." (cocoa = CŌCŌAH, plural of CŌA-TL); the sixth chapter, "which telleth of the various trees,…." ("quavitl" = CUAHUI-TL); the seventh chapter "which telleth of all the different herbs" ("xihujtl" = XIHUI-TL), including 150 types of medicinal herbs ("xihujtl patli"); and several concluding chapters that detail "the metals resting in the earth—tin, lead, and still others; and the different stones" (FC, p. 1).

This nicely defines the three domains: Animal, Vegetal, and Mineral, familiar to us as well, and indicates the major life-forms each includes. We will limit our purview here to the birds, the TŌTŌ-MEH.

The Náhuatl ornithological inventory begins with the lifeform "bird," TŌTŌ-TL. What makes a "bird" a "bird," for the Aztecs? They offer the following defining principle: "The property, the possession, which belongs to all the different birds and to turkeys is feathers <hivitl = IHHUI-TL>" (FC, p. 54).

The great majority of the birds described in the *Codex* are folk generic taxa (117 of 133, 88%), by Berlin's criteria. These include all those named binomially with TŌTŌ-TL as the head term, 25 by my count (see Appendix 3). Examples include QUETZAL-TŌTŌ-TL, the Resplendent Quetzal, and ILAMA-TŌTŌ-TL, the modest Canyon Towhee. These constitute approximately 21% of the 117 folk generic taxa named here. Other birds with two-part names that I treat as folk generic taxa include several "ducks," e.g., METZ-CANAUH-TLI, the Blue-winged Teal, and "eagles" such as MIX-CŌA-CUĀUH-TLI, the Ornate Hawk-Eagle. Though named binomially, in these cases the head term labels an intermediate taxon rather than a folk generic.

Folk specific taxa number approximately 16 (12%), each included within a named super-ordinate folk generic. Examples include the 11 kinds of hummingbirds (HUĪTZITZIL-IN) and birds such as the Great-tailed Grackle (TEŌ-TZANA-TL, "sacred grackle"), blackbirds and cowbirds (ĀCA-TZANA-TL, "reed grackle"), and the Slender-billed Grackle, called simply TZANA-TL, "grackle." There are no varietal taxa names and just three that I treat as intermediate categories: CANAUH-TLI$_1$ "duck," CUĀUH-TLI$_1$ "eagle," and TLOH-TLI$_1$ "falcon." Each of these three intermediate categories includes a prototypical folk generic taxon (or two in one case) that shares the name. These are distinguished by a subscripted number 2 (or 3): CANAUH-TLI$_2$ "Mexican Duck," CANAUH-TLI$_3$ "Muscovy Duck," CUĀUH-TLI$_2$ "Golden Eagle," and TLOH-TLI$_2$ "Prairie Falcon."

CANAUH-TLI$_1$ ("duck") includes a substantial fraction of the "waterfowl"—"*totome, at-lan nemj*," that is, "birds, living by the water"—listed in the third paragraph of Chapter 2 of Book 11 of the *Codex*. CUĀUH-TLI$_1$ "eagle," and TLOH-TLI$_1$ "falcon" are the primary examples of "birds of prey" treated in the fourth paragraph, collectively designated as "*tlavitequini*"—of uncertain derivation, possibly compounded of TLĀHUI-TL "red ochre" (Karttunen 1983:270), suggestive of blood, and TEQUI-TL "to cut something" (Karttunen 1983:232). These two explicitly named chapter headings, the "waterfowl" and the "birds of prey," are what Berlin treats as "special purpose categories," that is, categories motivated by considerations other than the formal properties of the birds themselves, which are said to define the "general purpose categories" of the formal folk taxonomy. In these cases, the defining feature is habitat or predatory behavior. These categories nonetheless are suggestive of broadly significant relationships amongst the local birds, as will be noted below.

How are Birds Named?

Bird names may be just that, the name of the bird, not allowing any further analysis. These names lack "descriptive force." In English a "duck" is just that. Likewise, we have "swallows," "robins," and "sparrows." Some Aztec bird names are similarly basic: ALO is the Scarlet Macaw; TZOPĪLŌ-TL is a vulture, AZTA-TL is the Snowy Egret. (The suffix -TL simply marks these as singular nouns.)

However, many bird names are onomatopoetic, that is, the name mimics a characteristic vocalization of the bird, like English "Killdeer." At least 24 (18%) of the birds listed in the *Codex* are so named (see Appendix 5). To facilitate this analysis, the Aztec scribes provided explanations for many of the names, perhaps at the request of Sahagún. For example, Ā-CHALALAC-TLI is the Belted Kingfisher and we are told that: "*Llámase por este nombre porque su canto es cha cha cha chu chu chala chala chala* <It is called by this name because its song is *cha cha cha chu chu chala chala chala* [author's translation]>" (García Quintana & López Austin, 1988, p. 703). Another example is HUĪLŌ-TL, Mourning Dove: "And it seems constantly to weep; it makes [the sound] *uilo-o-o*. And its name, *uilotl*, is taken from its song, which says *uilo*" (FC, p. 51) (nice approximation of the Mourning Dove's call).

Color terms play an important role in Aztec bird names. However, we need to be aware that color terms vary across languages (Berlin & Kay, 1969). The Aztec scribes employed a large number of color-related terminology to describe the appearance of the listed birds, including subtle shadings of red, yellow, blue, and green, some defined in terms of natural objects, such as "herb-green," "chili-red," "flower-yellow," the color of jade/turquoise, the color of the *tuna* (the ripe fruit of the prickly-pear cactus), and ochre. Of this welter of subtle shadings, the Aztec scribes applied 13 terms to name some 21 kinds of birds, which I list below (see also Appendix 4):

- IZTAC- "white," as in IZTAC-CUĂUH-TLI, "white eagle/hawk"; IZTAC-TLOH-TLI, "white falcon."
- YAUH-TIC, "dark," as in YAUH-TIC HUĬTZIL-IN, "dark hummingbird."
- CHĬL/CHIL- "red," as in CHĬL-TOTOPIL, "little red bird"; CHĬL-CANAUH-TLI, "red duck."
- AYO-PAL-TIC, "purple dye," as in AYOPAL-HUĬTZIL-IN, "purple hummingbird."
- TLA-PAL- "red ink," as in TLAPAL-HUĬTZIL-IN, "red hummingbird."
- COZ-TIC/COZCA-/ "yellow <or "necklace, ornament">", as in COZCA-CUĂUH-TLI, "yellow eagle/hawk"; COZ-TLOH-TLI, "yellow falcon."
- XOCHI-, "[color of] flower," as in XOCHI-TEN-ACAL, "flower beak."
- CUAPPACH-TIC/CUAPACH- "tawny," as in CUAPPACH-TŌTŌ-TL, "tawny bird"; CUAPPACH-HUĬTZIL-IN, "tawny hummingbird."
- NŌCH-, "[color of] prickly pear cactus fruit," as in NŌCH-TŌTŌ-TL, "cactus fruit red bird."
- CHALCHIUH-, "[color of] jade," CHALCHIUH-TŌTŌ-TL "jade bird"; CHALCHI-HUĬTZIL-IN, "jade hummingbird."
- XIHUA-PAL-/XIUH-PAL-, "by means of turquoise," as in XIHUA-PAL-QUECHŌL, "turquoise pendulum."
- XIHUI-, "light blue," as in XI-HUĬTZITZIL-IN, "light blue hummingbird."
- XIUH-TIC/XIUH- "[color of] jade, turquoise," XIUH-TŌTŌ-TL, "turquoise bird"; XIUH-QUECHŌL, "turquoise pendulum."

In addition to plumage colors, notable patterns might be named. The wind god, EHĒCA-TL, is characteristically portrayed in Aztec iconography with a "painted face." This facial pattern has motivated the naming of three quite different bird species, all showing the "painted face" of the wind god (Figure 36). The Wood Duck is ECA-TŌTŌ-TL, "wind god bird." "It is called *ecatototl* because its black feathers adorn the face [in the manner of the wind god]" (FC, p. 35). ECA-HUĬTZIL-IN, "wind god hummingbird," is actually two species, one with a pale face marked by a dark stripe through the eye, the other with a dark face marked by a white stripe through (or over) the eye. Though it is difficult to be certain, I believe the first is the Plain-capped Starthroat; the second is the White-eared Hummingbird. Both are common in the central highland of Mexico and fit the bill. Finally, ECA-TLOH-TLI, "falcon of the wind god," is, I believe, the Aplomado Falcon, at one time more widespread in the highlands north of the Valley of Mexico. It clearly shows this distinctive "wind god" facial ornament: "It is named *ecatotli* because the face is shot across with white" (FC, p. 44).

Aztec bird names often specified key habitat preferences and seasonal movements. Most often noted was an aquatic habitat, indicated by the prefix Ā-, "water." Examples, include Ā-CĀCĀLŌ-TL, literally, "water raven," for the White-faced Ibis; Ā-CHALALAC-TLI, for the Belted Kingfisher, named for its rattling vocalizations; Ā-CUĂUH-TLI, literally, "water eagle," for what might be the Crane Hawk and Ā-ITZ-CUĀUH-TLI for the Osprey, literally, "water obsidian eagle," that is, the "Golden Eagle of the water." We have Ā-TAPALCA-TL, literally, "pottery sherds of the water," for the Ruddy Duck, named for the sound of its wings beating on the water in the mating season. Ā-TEPONĀZ-TLI, literally, "log-drum of the water," for the American Bittern, descriptive of its booming mating calls. Ā-TŌTO-LIN, literally, "water turkey," names the American White Pelican, said to be "the ruler, the leader of all the water birds" (FC, p. 29). Ā-ZŌL-IN, "water quail," is the Wilson's Snipe, with its quail-like plumage. We also may note as a prefix ĀCA- "reed," to indicate habitat, as in, ĀCA-CHICHIC-TLI, for what I believe is the Least Bittern, named for its calls and its habitat. ĀCA-TZANA-TL, literally, "reed grackle," for blackbirds and cowbirds in winter flocks.

CUĀUH-TLI is the general term for an eagle or large hawk. However, as a prefix, CU-AUH-, with a short A, may indicate something wild or "of the forest," cf. CUAHUI-TL, "tree, wood" (Karttunen, 1983, p. 58). For example, CUAUH ALO, "forest/wild macaw," names the Military Macaw, in contrast to ALO, the Scarlet Macaw. The Crested Guan is called CUAUH-TŌTO-LI, literally, "forest/wild turkey." The prototypical woodpecker, likely the Golden-fronted, is CUAUH-TOTOPO-TLI, literally, "strikes wood (with sound of thunder)."

HUEXO-CANAUH-TLI, "willow duck," is, I believe, the Green Heron, based on other descriptive details. A willow habitat is quite appropriate for this "duck." (The Aztecs also called the Black-crowned Night-Heron and the Belted Kingfisher "ducks.") XOPAN-TŌTO-TL, literally, "summer bird," is another name for the Lesser Goldfinch, also known as MIYAHUA-TŌTO-TL, "maize tassel bird," suggestive of the seasonal gathering of these colorful birds as the maize crop flowers. CENOTZ-QUI, literally, "brings frost," one of several names for the Merlin, is suggestive of its status as a winter visitor. YOHUAL-TLOH-TLI, "nocturnal falcon," might name the Bat Falcon, known for its crepuscular pursuit of bats. ZACA-TECOLŌ-TL, literally, "grass owl," is likely the Striped Owl or the Short-eared Owl, both partial to grasslands, while ZACA-TLA-TLI, "grass eater," is a general term for sparrows. ZOQUI-Ā-ZŌL-IN, "mud quail," is an elaboration of Ā-ZŌL-IN, "water quail," to name the Wilson's Snipe. XĀL-CUĀ-NI, literally, "eats sand," is the American Wigeon, which often forages along sandy shores, but in search of water plants to eat.

Finally, Aztec bird names may describe the shape or color of particular body parts, as in ICXI-XOXOUHQUI, "has green legs," for the American Avocet, or TEN-ITZ-TLI, literally, "obsidian beak," for some sort of "nocturnal" water bird, which I suspect might have been the Black Tern. YACA-CIN-TLI, "corn-cob nose," is the American Coot; YACA-PATLA-HUAC, "flat-nose," is the Northern Shoveler, and YACA-TEXO-TLI, "blue-nose," is another name for the Ruddy Duck. The male in breeding season sports a bright powder-blue beak.

These examples demonstrate the value of close attention to the "descriptive force" of "arbitrary" names for things.

What's Missing?

The birds detailed in the *Florentine Codex* provide an impressive introduction to Aztec ethno-ornithology, a systematic inventory of over 130 distinct, named kinds of birds with fascinating details closely observed and commentary on their roles in Aztec lives. However, those of us obsessed with our feathered fauna spot puzzling gaps in the *Codex* accounts. There are a number of birds the Aztec scribes might well have encountered in their home territory that are sufficiently conspicuous from our perspective that we may wonder why they were passed over or perhaps treated in an off-hand manner. For example, why are the Black and Turkey Vultures not more clearly differentiated? Though both are scavengers, they differ in appearance, flight patterns, and foraging strategies. Raptors are particularly diverse in the central Mexican highlands and several distinctive species seem to have been overlooked, such as the Harris' Hawk (*Parabuteo unicinctus*) with its distinctive cooperative hunting strategy and the ubiquitous Red-tailed Hawk (*Buteo jamaicensis*). The various quail and wood-partridge are subsumed within a single folk generic category, for which the shy, though endearing, Montezuma Quail serves as the prototype. What became of the roadrunners, our favorite cartoon character? Two species are common and widespread in the Aztec homeland (*Geococcyx velox* and *G. californianus*), though neither is named (but see Appendix 1).

It is understandable that the many species of migrant shorebirds, most here in their subtle gray winter feathers, should be "lumped" as TEZOLOC-TLI. However, we might have expected the Killdeer (*Charadrius vociferus*), described as "widespread, noisy, and conspicuous" (Howell & Webb, 1995, p. 257), to have been singled out (but see Appendix 1). The Great Horned Owl was certainly a familiar nocturnal voice, but there is no indication that this or any other owl species was feared as a harbinger of misfortune or death, as so many of the world's rural people forfend. Two swallow species are distinguished, both mud-nesters and both therefore familiar about town. However, what of the swifts (Apodidae)—masters of the skies and weather forecasters? Tyrant flycatchers (Tyrannidae) are notably absent from the *Codex* accounts, with the sole exception of the Vermilion Flycatcher. In my experience this family is well represented by such in-your-face species as the kingbirds (*Tyrannus*), the raucous Great Kiskadee (*Pitangus sulphuratus*), and Bright-rumped Attila (*Attila spadiceus*).

One might have expected the Steller's Jay (*Cyanocitta stelleri*) or another of its colorful cousins to have been duly noted, but there appears to be no record of them. The scarcity of names for the diminuitive woodland species, the vireos (Vireonidae) and wood-warblers (Parulinae), is not surprising, but I still missed the American Dipper (*Cinclus mexicanus*) that patrols the mountain streams, the flashy Gray Silky-Flycatchers (*Ptilogonys cinereus*), and the Northern Cardinal (*Cardinalis cardinalis*). Several of these "missing" species are named in more contemporary regional inventories, as we list in Appendix 1.

And finally, whatever happened to the bats (Chiroptera)? Given that the Aztecs defined birds as creatures with feathers, one would not expect them to conflate bats with birds, as is common in other cultural traditions. However, bats were certainly well known to the Aztecs as TZINĀCAN-TLI but are not listed anywhere in Book 11. In fact, the Aztec merchants, the Pochteca, travelled in disguise to the Mayan kingdom of Tzinacantan (TZINĀCAN-TLAN, literally, "Land of Bats") in present-day Chiapas state to surreptitiously acquire precious quetzal plumes. Perhaps bats also "fell through the cracks," being neither fish nor fowl.

Conclusions

The Aztecs in the eyes of many Western academics were a rather barbaric bunch, hardly worthy of the title "civilized." Their large-scale human sacrifice of war captives is a dominant trope, as their priests ripped the hearts from the chests of sacrificial victims and then rolled their bodies down the pyramid steps. Aztec iconography may be grotesque and may repel at first glance. However, there is clearly another side to the Aztec portrait. The elegant poetry of the Texcocan ruler, Nezahualcoyotl, testifies to a deep appreciation for the beauty of the natural world. Aztecs treated flowers as "prayers," offerings perhaps on a par with the blood of human sacrifices. The Aztec fascination with their diverse avian companions offers further support for the view of the Aztecs as our intellectual and moral equals.

Sahagún came to value the Aztec people as fully human, despite the cruelty of their ritual offerings to the Sun, the god that demanded blood sacrifice. "What is certain is that all these peoples are our brothers, proceeding from Adam's stock even as we ourselves, they are our neighbors whom we must love as ourselves" (from the Prologue to the *Historia General* as quoted in Todorov [1984, p. 239]; Spanish original in García Quintana & López Austin, 1988, Volume 1, p. 35). His account of the Aztec practice of child sacrifice is deeply empathetic:

> I do not believe there can be a heart so hard as not to be touched and softened by tears, horror and fear, in hearing of cruelty so inhuman,…. The cause of this cruel blinding, of which these wretched children were the object, must not be so much imputed to the cruelty of their fathers, who were shedding abundant tears and indulging in such practices with pain in their hearts; we must impute it to the infinite cruel hatred of our ancient enemy Satan (from the *Historia General* as quoted in Todorov [1984, p. 228]; Spanish original in García Quintana & López Austin, 1988, Volume 1, pp. 106–107).

Sahagún is torn between his allegiance to his Christian ideals and his deep empathetic identification with his Aztec students. "Sahagún's explicit project was to facilitate the evangelization of the Indians by the study of their religion. But barely a third of the book corresponds to this idea" (Todoov, 1984, p. 235). "But alongside this declared motive there exists another … the desire to know and to preserve Nahuatl culture" (Todorov, 1984, p. 223). "The result of these efforts is an inestimable encyclopedia of the spiritual and material life of the

Aztecs before the conquest" (Todorov, 1984. p. 227). The conquered speak through this agent of conquest.

Sahagún's ethnography is an enduring model for the best practice of cultural anthropology today. Sahagún does not interpret Native voices. Rather, he lets them speak for themselves in their Native tongue. This, I believe, is the proper task of ethnography, then and now. I believe in an anthropology that first and foremost seeks to honor our common humanity by bearing witness to all that we may share as citizens of a transhuman society of nature.

Appendix 1

Náhuatl Bird Names from Other Sources

Náhuatl, also known as Mexicano, is spoken by some two million in Mexico today. Variant Náhuatl bird names have been recorded at various times and places. Listed here are bird names from Tetelcingo, Xalitla, and Zacapoaxtla, cited in Karttunen (1983) and from Molina, cited in Wood (2000–2020), that are not listed in the *Florentine Codex*. Names for plants and animals readily shift their ornithological referents through time and space, as each speech community applies a stock of names to a varying local flora and fauna.

- Ā-CĀHUA-TŌTŌ-TL, "small yellow or blue bird with gray on back" (Tetelcingo in Karttunen, 1983, p. 1).
- Ā-HUACĀ-TZIN, "swallow" (Tetelcingo in Karttunen, 1983, p. 7).
- AMACOZCA-TL. Killdeer (*Charadrius vociferus*) (Friedmann et al., 1950, p. 91).
- Ā-TŌTO-TL, "heron" (IDIEZ [Wood, 2000–2020]).
- Ā-XOCXOC-TZIN, "a type of bird with green plumage" (Xalitla in Karttunen, 1983, p. 15).
- CĀCALĀCH-TŌTŌ-TL, literally, "cockroach bird," "Mexican flycatcher / *avispero*," possibly Tyrannidae (Tetelcingo in Karttunen, 1983, p. 19). This could refer to a large tyrant flycatcher (Tyrannidae), of which there are numerous possibilities.
- CANAUH CACUAN-TŌTŌ-TL, Cedar Waxwing (*Bombycilla cedrorum*) (Miller et al., 1957, p. 211).
- CANAUH-CONE-TON-TLI, "small duck species" (Molina in Wood, 2000–2020).
- CHIAN-TŌTŌ-TL, Black-throated Sparrow (*Amphispiza bilineata*) (Miller et al., 1957, p. 379).
- CHIHUIHCOYŌ, "partridge, *perdiz*" (Zacapoaxtla in Karttunen, 1983, p. 54); perhaps refers to the Long-tailed Wood-Partridge (*Dendrortyx macroura*) or a tinamou (Tinamidae).
- CHĪL-TŌTŌ-TL, literally "red bird," "cardinal" (Tetelcingo in Karttunen, 1983, p. 52); perhaps distinguished from CHĪL-TOTOPIL by larger size, in which case it could well be the Northern Cardinal (*Cardinalis cardinalis*), another conspicuous bright red bird.

- CHOHCHOC, "a type of bird / *pajaro arriero* <muleteer bird>," possibly a roadrunner (Zacapoaxtla in Karttunen, 1983, p. 54); if this might refer to the roadrunner, two species are possible, Greater (*Geococcyx californianus*) and Lesser (*G. velox*).
- COXCOX, "coffee color, come out at night on the roads" (IDIEZ: nightjar).
- COXCOX-TLI/COXOLI-TLI, "type of pheasant" (Molina in Wood, 2000–2020); as there were no "pheasants" in 16th-century Mexico, I suspect this term may have referred to some species of tinamou (Tinamidae), of which the Thicket Tinamou (*Crypturellus cinnamomeus*) is the most likely, as it ranges further north on both coasts (Howell & Webb, 1995, p. 89–91). Its coloration also may be described as "coffee colored" (COXCOX). See also IXMATLA-TŌTŌ-TL (perhaps Great Tinamou) and TECUZIL-TŌTŌ-TL (perhaps Thicket Tinamou). COXOLI-TLI "type of pheasant," Crested Guan (Friedmann et al., 1950).
- COZTIC-TŌTŌ-TL, literally, "yellow bird," "oriole / *calandria*" (Tetelcingo in Karttunen, 1983, p. 43); there are several species of orioles (*Icterus* species) within the range of the Aztecs.
- CUAUH ALO, literally, "forest macaw," Military Macaw (*Ara militaris*) "large green parrot" (Molina in Wood, 2000–2020); though not mentioned in the *Codex*, this species is distinguished from ALO, the Scarlet Macaw (*Ara macao*).
- CUAUH-QUEQUĒX, "woodpecker species": (Zacapoaxtla in Karttunen, 1983, p. 63). See also CUAUH-TOTOPOC-TLI, Golden-fronted Woodpecker (*Melanerpes aurifrons*).
- CUAUH-TZIC-TŌTŌ-TL, "a type of oriole that eats sap … *avispero*" (Tetelzingo in Karttunen, 1983, p. 65); though in no way resembling an oriole, the Yellow-bellied Sapsucker (*Sphyrapicus varius*)—a rather aberrant woodpecker and regular winter visitor throughout Mexico—is noted for that feeding strategy.
- CUAUH-XAXACA, "a type of horned owl / *tecolote, buho*" (Zacapoaxtla in Karttunen, 1983, p. 65).
- CUÏX-IN, "large bird of prey, hawk, *milano, gavilan*" (Xalitla in Karttunen, 1983, p. 74). See also CUĀUH-TLI "eagle/hawk in general." Not to be confused with CAHUÏX-IN or COHUÏX-IN.
- HUAU-TŌTŌ-TL, Painted Bunting (*Passerina ciris*) (Miller et al., 1957, p. 338).
- HUĒI-TŌTO-LIN, literally, "big turkey," Wild Turkey (*Meleagris gallopavo*), "a type of wild bird, probably the wild turkey / *tortola silvestre*" (Zacapoaxtla in Karttunen, 1983, p. 86); see HUEHXŌLŌ-TL.
- ĪXTĒN-TLI MŌYŌ-TZIN, Yellow-eyed Junco (*Junco phaenotus*) (Miller et al., 1957, p. 385).
- IZTAC-[T]ZONYAYAUHQUI, Wood Duck (*Aix sponsa*) (Friedmann et al., 1950, p. 43).
- QUICHPĀYA-TL, "type of bird, *agarista*" (Tetelcingo in Karttunen, 1983, p. 210).
- QUILTOTON, Mexican Parrotlet (*Forpus cyanopygius*) (Friedmann et al., 1950, p. 127).

- TEXIXĪCTŌN, "Cañon Wren (*Catherpes mexicanus*)" (Tetelcingo in Karttunen, 1983, p. 238); see TLATHUICICI-TLI.
- TEXOLOC-TLI, Ring-necked Duck (*Aythya collaris*) (Friedmann et al., 1950, p. 43).
- TLĀCA AZTA-TL, "heron / egret species" (Wood, 2000–2020). This might apply to other egrets than the Snowy, AZTA-TL, such as the Great Egret (*Egretta alba*).
- TLĀLA-CATE-TL, "dove / *paloma*" (Xalitsa in Karttunen, 1983, p. 273).
- TLAŌL-TŌTŌTL, "type of bird, *maizero*" (Tetelcingo in Karttunen, 1983, p. 287).
- TOLOLOH, "owl / *tecolote, buho*" (Zacapoaxtla in Karttunen, 1983, p. 244).
- XAXACA, "owl / *tecolote, buho*" (Zacapoaxtla in Karttunen, 1983, p. 322). See also CUAUH-XAXACA.
- YOHUAL-TŌTŌ-TL, literally, "nocturnal bird" (Wood, 2000–2020). An owl (Strigiformes) or perhaps a nightjar (Caprimulgidae) of some kind.
- ZACA-PIPĪTZTŌN, "a type of bird" (Tetelcingo in Kartunnen, 1983, p. 345).

Appendix 2

Taxonomic Structure as Shown in the Nomenclature or in the Paragraph Headings

Major groupings specifically named in the paragraph headings:
- TŌTŌ-MEH, Ā-TLAN NEMI "waterfowl"
- IXQUICHT-IN TOTO-ME, TLAHUITE-QUI-NI "birds of prey"
- TŌTŌ-MEH, HUEL CUĪCA-NI "birds which are good singers"

Intermediate categories that include numerous folk generics, as listed following each heading:
- CANAUH-TLI ("duck") (*CANAUH-TLI absent in the name, 13 of 20 species)
  - *Ā-CHALALAC-TLI (Belted Kingfisher)
  - *AMANACOCHE (Bufflehead)
  - *Ā-TAPALCA-TL (Ruddy Duck)
  - CANAUH-TLI$_2$ (Mexican Duck)
  - *CUA-CHIL-LI/CUA-CHIL-TON (Common Gallinule)
  - CHIL-CANAUH-TLI (Cinnamon Teal)
  - CON-CANAUH-TLI (Fulvous Whistling-Duck)
  - *CUA-COZ-TLI (Canvasback)
  - *ECA-TŌTŌ-TL (Wood Duck)
  - *HUĀC-TLI/HUAC-TLI$_1$ (Black-crowned Night-Heron)
  - HUEXO-CANAUH-TLI (Green Heron)
  - METZ-CANAUH-TLI (Blue-winged Teal)
  - *QUETZAL-TEZOLOC-TON (Green-winged Teal)
  - *TLALALACA-TL (Black-bellied Whistling-Duck)
  - *TZITZIHUA (Northern Pintail)
  - *TZON-YAYAUHQUI (Lesser Scaup)
  - *XALQUANI (American Wigeon)
  - *YACA-PATLA-HUAC (Northern Shoveler)
  - ZŌL-CANAUH-TLI (Gadwall)
  - ZOQUI-CANAUH-TLI (Fulvous Whistling-Duck)

- CUĀUH-TLI₁ ("eagle")
  - Ā-CUĀUH-TLI (possibly Crane Hawk)
  - Ā-ITZ-CUĀUH-TLI (Osprey)
  - CHIYAN-CUĀUH-TLI (Northern Harrier)
  - COZCA-CUĀUH-TLI (Crested Caracara)
  - ITZ-CUĀUH-TLI/CUĀUH-TLI₂ (Golden Eagle)
  - IZTAC-CUĀUH-TLI (possibly White-tailed Hawk)
  - MIX-CŌĀ-CUĀUH-TLI (Ornate Hawk-Eagle)
  - TLĀCO-CUĀUH-TLI (Northern Harrier)
  - TLOH-CUĀUH-TLI (Northern Goshawk)
  - YOHUAL-CUĀUH-TLI ("nocturnal eagle," possibly Northern Potoo)
- TLOH-TLI₁/TLO-TLI₁ (falcon) (note several apparent synonyms)
  - ĀYAUH-TLOH-TLI "cloud falcon" (Peregrine Falcon)
  - COZ-TLOH-TLI "yellow falcon" (American Kestrel)
  - CUĀUH-TLOH-TLI "eagle falcon" (Peregrine Falcon)
  - ECA-TLOH-TLI (Aplomado Falcon)
  - ITZ-TLOH-TLI/ĀCA-TLOH-TLI "obsidian falcon"/"reed falcon" (Bat Falcon)
  - IZTAC-TLOH-TLI "white falcon" (Prairie Falcon)
  - *NECUILIC-TLI/NECUILOC-TLI/CENOTZQUI/TLETLEUHTON (Merlin)
  - TLĀCO-TLOH-TLI (Northern Goshawk)
  - TLOH-TLI₂/TLO-TLI₂ (Prairie Falcon, as prototype)
  - YOHUAL-TLOH-TLI (Bat Falcon)

Polytypic folk generic taxa with included folk specifics listed following (*non-binomial names):
- HUĪTZITZIL-IN/HUĪTZIL-IN (hummingbird)
  - AYOPAL-HUĪTZIL-IN (perhaps the Violet-crowned Hummingbird)
  - CHALCHI-HUĪTZIL-IN (possibly the Mexican Violetear)
  - CUAPPACH-HUĪTZIL-IN (likely the Cinnamon Hummingbird)
  - ECA-HUĪTZIL-IN (Plain-capped Starthroat and White-eared Hummingbird)
  - QUETZAL-HUĪTZIL-IN (perhaps the Garnet-throated Hummingbird)
  - TELOLO-HUĪTZIL-IN (perhaps a category including females and immatures of several species)
  - TLAPAL-HUĪTZIL-IN (likely the Rufous Hummingbird)
  - TLE-HUĪTZIL-IN (species indeterminate)
  - *TOTOZCATLE-TON (Broad-tailed or Lucifer Hummingbird)
  - XI-HUĪTZITZIL-IN (perhaps the Broad-billed Hummingbird)
  - YAUHTIC HUĪTZIL-IN (species indeterminate)
- ALO (macaw)
  - ALO (Scarlet Macaw)

- ◦ CUAUH ALO "forest macaw" (Military Macaw)
- AZTA-TL (egret)
  - ◦ AZTA-TL/TEO-AZTA-TL (Snowy Egret)
  - ◦ TLĀCA-AZTA-TL (possibly the Great Egret)
- CĀCĀLŌ-TL
  - ◦ CĀCĀLŌ-TL (Common Raven)
  - ◦ Ā-CĀCĀLŌ-TL "water raven" (White-faced Ibis)
- CHĪCUA-TLI
  - ◦ CHĪCUA-TLI (Barn Owl)
  - ◦ TLĀL-CHĪCUA-TLI "ground barn owl" (Burrowing Owl)
- HUĪLŌ-TL (dove)
  - ◦ HUĪLŌ-TL (Mourning Dove)
  - ◦ TLĀCA-HUILO-TL (Rock Pigeon)
- QUECHŌL-LI
  - ◦ QUECHŌL-LI/TLĀUH-QUECHŌL-LI/TLĀUH-QUECHŌL (Roseate Spoonbill)
  - ◦ XIHUA-PAL-QUECHŌL/XIUH-PAL-QUECHŌL (Turquoise-browed Motmot)
  - ◦ XIUH-QUECHŌL-LI/XIUH-QUECHŌL (Lesson's Motmot)
- TECOLŌ-TL (owl) (**from sources other than the *Florentine Codex*)
  - ◦ **COPĀC-TECOLŌ-TL (American Kestrel)
  - ◦ TECOLŌ-TL (Great Horned Owl)
  - ◦ **TŌCH-TECOLŌ-TL (Band-tailed Pigeon)
  - ◦ ZACA-TECOLO-TL "grass owl" (Striped or Short-eared Owl)
- TŌTO-LIN (turkey-like birds)
  - ◦ CUAUH-TŌTO-LIN (Wild Turkey)
  - ◦ *HUIĒ-TŌTO-LIN
  - ◦ TŌTO-LIN/TŌTO-LI (domestic turkey)
    - · HUEHXŌLŌ-TL (male turkey)
  - ◦ Ā-TŌTO-LIN (American White Pelican)
- TZANA-TL (grackles and allies)
  - ◦ ĀCA-TZANA-TL "reed grackle" (blackbirds/cowbirds)
  - ◦ TEŌ-TZANA-TL "sacred grackle" (Great-tailed Grackle)
  - ◦ TZANA-TL (Slender-billed Grackle)
- ZŌL-IN ("quail")
  - ◦ Ā-ZŌL-IN/ZOQUI-Ā-ZŌL-IN "water quail"/"mud quail" (Wilson's Snipe)
  - ◦ ZŌL-IN/ZOL-LI (Montezuma Quail)
    - · HUATON (female Montezuma Quail)
    - · TECU-ZOLI (male Montezuma Quail)

Appendix 3

Bird Names Incorporating
TŌTŌ-TL/TŌTŌ-MEH (Bird/Birds)

TŌTŌ-TL/TŌTŌ-MEH (bird/birds): "The property, the possession, which belongs to all the different birds and to turkeys is feathers (***ihhui-tl***)" (FC, p. 54).

TŌTŌ-TL (*if not listed in the *Codex* but from other sources, e.g., Friedmann et al., 1950; Karttunen, 1983; Wood, 2000–2020)

- *Ā-CĀHUA-TŌTŌ-TL ("small yellow and blue bird with gray back")
- *Ā-TŌTŌ-TL (egret)
- Ā-YACACH-TŌTŌ-TL (*Campylorhynchus* wrens)
- *CĀCALĀCH-TŌTŌ-TL (flycatcher species)
- *CAPOTZTIC-TOTO-TL (blackbirds)
- CHIUH-TŌTŌ-TL (Red-legged Honeycreeper)
- CHĬCUA-TŌTŌ-TL (Eastern Meadowlark)
- CHĬL-TOTOPIL (Red Warbler)
- *CHĬL-TŌTŌ-TL (Northern Cardinal)
- COYOL-TŌTŌ-TL (Red-winged/Yellow-headed Blackbirds)
- COZTIC-TŌTŌ-TL (oriole)
- CUAPPACH-TŌTŌ-TL/CUAPACH-TŌTŌ-TL (Squirrel Cuckoo)
- CUAUH-TŌTO-LIN (Wild Turkey)
- *CUAUH-TZIC-TŌTŌ-TL (oriole/sapsucker)
- CUĬTLACOCH-TŌTŌ-TL/CUĬTLACOCH-INQUI (Curve-billed Thrasher)
- ECA-TŌTŌ-TL (Wood Duck)
- ĒLŌ-TŌTŌ-TL (Blue Grosbeak)
- ILAMA-TŌTŌ-TL (Canyon Towhee)
- *ĪXCAUH-TŌTŌ-TL (Groove-billed Ani)
- IXMATLA-TŌTŌ-TL (Great Tinamou)
- MIYAHUA-TŌTŌ-TL (Lesser Goldfinch)
- MOLO-TŌTŌ-TL/MOLO-TL (House Finch)

- NŌCH-TŌTŌ-TL (male House Finch)
- QUETZAL-TŌTŌ-TL (Resplendent Quetzal)
- *QUIYAUH-TŌTŌ-TL ("rain-storm bird")
- TECUZIL-TŌTŌ-TL (Thicket Tinamou)
- *TLAŌL-TŌTŌ-TL ("maizero")
- TLAPAL-TŌTŌ-TL (Vermilion Flycatcher)
- XIUH-TŌTŌ-TL (Lovely Cotinga)
- XOCHI-TŌTŌ-TL (Black-backed Oriole)
- XOPAN-TŌTŌ-TL (Lesser Goldfinch)
- *YOHUAL-TŌTŌ-TL (nocturnal bird)
- YOLLO-TŌTŌ-TL (Rose-throated Becard)
- ZA-CUAN-TŌTŌ-TL/ZA-CUAN (Montezuma Oropendola)

Appendix 4

Some Modifying Attributives

- Ā- "water" [Ā-TŌTO-LIN, Ā-COYO-TL, Ā-CIH-TLI, Ā-TAPALCA-TL, Ā-CHACHALAC-TLI, Ā-CACHICHIC-TLI, Ā-CUĀUH-TLI, Ā-ITZ-CUĀUH-TLI, Ā-CACALO-TL, Ā-ZŌL-IN]
- ĀCA- "reed" [ĀCA-CHICHIC-TLI, ĀCA-TLOH-TLI, ĀCA-TZANA-TL]
- ĀYAUH- "cloud" [ĀYAUH-TLOH-TLI]
- AYOPAL- "purple," [AYOPAL-HUĪTZIL-IN]
- CENTZON-/ZENTZON- "many [400]" [CENTZON-/ZENTZON-TLAHTŌLEH]
- CHALCHIUH- "turquoise" [CHALCHIUH-TŌTŌ-TL, CHALCHI-HUĪTZIL-IN]
- CHICUA- "barn owl" [CHICUA-TŌTŌ-TL]
- CHĬL/CHIL- "red" [CUA-CHICHIL, CUA-CHIL-TON, CHĬL-TOTOPIL, CHIL-CANAUH-TLI]
- CON- "pot" [CON-CANAUH-TLI]
- COYOL- "rattle, bell" [COYOL-TŌTŌ-TL]
- COZCA-/COZTIC-/COZ- "yellow"; "ornament, necklace" [CUA-COZ-TLI, COZCA-CUĀUH-TLI, COZ-TLOH-TLI]
- CUAPPACH-/CUAPACH- "tawny [*leonado*]" (cf. -PACH- > "plant refuse") [CUAPPACH-TŌTŌ-TL, CUAPPACH-HUĪTZIL-IN]
- CUĀ-/CUAH- "to eat something' [XAL-CUA-NI]
- CUAHUI-TL "tree, wood, stick, staff, beam" [CUAUH-TŌTO-LI, CUAUH-TOTOPO-TLI]
- CUĀUH- "eagle" [CUĀUH-TOH-TLI]
- ECA- "wind god" [ECA-TŌTŌ-TL, ECA-HUĪTZIL-IN, ECA-TLOH-TLI]
- HUEXO- "willow" [HUEXO-CANAUH-TLI]
- ILAMA- "old woman" [ILAMA-TŌTŌ-TL]
- ITZ- "obsidian" [ITZ- CUĀUH-TLI, TEN-ITZ-TLI, ITZ-TLOH-TLI]
- IZTAC- "white" [IZTAC-CUĀUH-TLI, IZTAC-TLOH-TLI]
- METZ- "moon" [METZ-CANAUH-TLI]
- MIYAHUA- "maize tassel" [MIYAHUA-TŌTŌ-TL]
- NŌCH- "prickly pear cactus fruit ["tuna"]" [NŌCH-TŌTŌ-TL]

- QUECHŌL- "pendulum motion" [TLĀUH-QUECHŌL, TEŌ-QUECHŌL, XIUH-QUECHŌL, XIHUA-PAL-QUECHŌL]
- QUETZAL- "resplendent" [QUETZAL-TŌTŌ-TL, QUETZAL-TEZOLOC-TON, QUETZAL-HUĬTZIL-IN]
- TELOLO- "ball" [TELOLO-HUĬTZIL-IN]
- TEN- "nose/beak" [XOCHI-TEN-ACAL, TEN-ITZ-TLI]
- TEŌ- "sacred" [TEŌ-QUETZAL, TEŌ-QUECHŌL, TEŌ-TZINITZCAN, TEŌ-TZANA-TL]
- TLĀCA-/TLACA- "person"; or "respectful form of address" [TLĀCA-HUILO-TL]
- TLAHCO- "middle, center," perhaps "typical"; TLACŌ-TL "staff, stick, switch" [TLAHCO-CUĀUH-TLI]
- TLĀL- "earth, land, property" [TLĀL-CHICUA-TLI]
- TLAPAL- "red dye, red ink," "something dyed/blood" [TLAPAL-HUĬTZIL-IN, TLAPAL-TŌTŌ-TL]
- TLĀUH- > TLĀHUI-TL "red ochre" [TLĀUH-QUECHŌL]
- TLOH- "falcon" [TLOH- CUĀUH-TLI]
- TZINITZCAN- "precious feathers of a bird," green or red
- XIUH-/XIHUA-/XI--, "jade" [XIUH-QUECHŌL, XIHUA-PAL-QUECHŌL, XI-HUĬTZITZIL-IN, XIUH-TŌTŌ-TL]
- XOCHI- "flower" [XOCHI-TEN-ACAL, XOCHI-TŌTŌ-TL]
- XOPAN- "springtime" [XOPAN-TŌTŌ-TL]
- YACA- "beak" [YACA-PITZA-HUAC, YACA-CIN-TLI, YACA-PATLA-HUAC]
- YAUH-TIC, "dark," [YAUHTIC HUĬTZIL-IN]
- YOHUAL- "nocturnal" [YOHUAL- CUĀUH-TLI, YOHUAL-TOH-TLI]
- YOLLO- "heart" [YOLLO-TŌTŌ-TL]
- ZACA- "grass" [ZACA-TECOLO-TL, ZACA-TLA-TLI]
- ZOQUI- "mud" [ZOQUI-CANAUH-TLI, ZOQUI-Ā-ZŌL-IN]

APPENDIX 5

Onomatopoetic or Sound Symbolic Names (N = 24)

Translated from Spanish by the author (all quotes are from García Quintana & López Austin, 1988). I include also birds named for some aspect of the characteristic vocalizations in cases not strictly-speaking onomatopoeia:

ĀCA-CHICHIC-TLI (Least Bittern) "Y llámase así porque su canto es *achichichic* <And it is named thus because its song is *achichichic*>" (p. 704).

Ā-CHALALAC-TLI (Belted Kingfisher) "Llámase por este nombre porque su canto es *cha cha cha chu chu chala chala chala* <It is called by this name because its song is *cha cha cha chu chu chala chala chala*>" (p. 703) (sound clip by Liza Verkony, ML436910561).

Ā-TAPALCA-TL (Ruddy Duck) "Llámanse *atapálcatl* porque, cuando quiere llover, un día antes y toda la noche hace ruido en el agua con las alas, batiendo en el agua con las alas <They are called *atapálcatl* because, when it wants to rain, a day before and throughout the night it makes noise in the water with its wings, whipping the water with the wings>" (p. 702). Not exactly onomatopoetic, but sound symbolic.

Ā-TEPONĀZ-TLI/Ā-TONCUEPO-TLI/TOL-COMOC-TLI (American Bittern)

> … for this reason is it called *tolcomoctli*: as it sings, it resounds. For this reason it is called *atoncuepotli*: when it sings, it is clearly heard to explode; it is very loud. And it is called *ateponaztli* because it sounds from a distance like a two-toned drum, so loud it is (FC, p. 33).

These three synonyms all specify the resemblance of the "booming" calls of the bittern to various loud noises familiar to the locals.

ĀYACACH-TŌTŌ-TL (*Campylorhynchus* wrens) "Llámase *ayacachtototl* porque canta como soena las sonajas que llaman *ayacachtli* [rattles]. Dice *chacha cha, xi xi xi xi, cha xe xi, cha xe chi, cho cho cho cho* <It is called *ayacachtototl* because when it sings it sounds like the rattles they call *ayacachtli*. It says *chacha cha, xi xi xi xi, cha xe xi, cha xe chi, cho cho cho cho*>" (p. 708).

CHACHALACAME-TL (West Mexican Chachalaca) "Canta en verano, y por eso le llaman *chachalacámetl*. Cuando se juntan muchas destas aves, una dellas comienza a cantra, y luego la sigen todas las otras <It sings in summer, and therefore they call it *chachalacámetl*. When many of these birds gather together, one of them begins to sing, and then all the others follow it>" (p. 712). The name is obviously onomatopoetic, though not explicit here.

CHIQUIMO-LIN (Imperial Woodpecker)

> And when it sings, it cries out much, it warbles, sometimes like whistling with the fingers; and it sings as if there were many birds…. But when it seems to shriek, it is angry…. But when it whistled, they said it was happy…. (FC, pp. 52–53).

Perhaps not really sound symbolic, but the name seems to reflect the vocalizations central to local perceptions of this species.

CŌCOH-TLI (Inca Dove) "Llámanse *cocotli* porque cuando cantan dicen *coco coco*…. No se casan más de una vez. Y cuando muere el uno, el otro siempre anda como llorando, y solitario, diciendo *coco coco* <They are called *cocotli* because when they sing they say *coco coco*…. They never marry more than once. And when one dies, the other always walks about as if crying, and alone, saying *coco coco*>" (p. 709).

COHUIX-IN (Aztec Rail) "Y llámanla ansí porque cuando canta dice *cohuixi, cóhuix* <And they name it thus because when it sings it says *cohuixi, cóhuix*>" (p. 700) (sound clip by Anuar López, ML432093161).

COYOL-TŌTŌ-TL (Red-winged and Yellow-headed Blackbirds) "Y cantan muy bien. Por esto se llama *coyoltototl*, que quiere decir 'ave que canta como cascabel' <And they sing very well. For this it is called *coyoltototl*, which means 'bird that sings like a rattlesnake'>" (p. 711).

CUĪTLACOCH-IN/CUĪTLACOCH-TŌTŌ-TL (Curve-billed Thrasher) "Canta muy bien. Llámase *cuitlacochtototl* por razón de su canto, que dice *cuitlácoch, cuitlácoch, tarata tarat tatatati*, etcétera <It sings very well. It is called *cuitlacochtototl* because of its song, that says *cuitlácoch, cuitlácoch, tarata tarat tatatati*, etcetera>" (p. 711) (sound clip by Andrew Theus, ML436112051).

HUĀC-TLI₁ (Black-crowned Night-Heron) "Llámase por este nombre, *huactli*, porque cuando canta dice *huac, huac* <It is called by this name, *huactli*, because when it sings it says *huac, huac*>" (p. 704).

HUĀC-TLI₂ (Laughing Falcon)

> It sings in this manner: sometimes it laughs like some man; like a man speaking it can pronounce these words: *yeccan, yeccan, yeccan*. When it laughs, it says *ha ha ha ha ha, ha hay, ha hay, hay hay, ay*. Especially when it finds its food it really laughs (FC, p. 42) (sound clip by Adam Betuel, ML436679641).

HUĪLŌ-TL (Mourning Dove) "And it seems constantly to weep; it makes [the sound] *uilo-o-o*. And its name, *uilotl*, is taken from its song, which says *uilo*" (FC, p. 51). A nice approximation of the Mourning Dove's call.

HUĪTZITZIL-IN (hummingbird). Not explicitly noted as onomatopoetic, but clearly so.

IXMATLA-TŌTŌ-TL (Great Tinamou) "Llámase por este nombre, *ixmatlatótotl*, porque su canto es como habla de persona. Dice cuando canta *campa huee, campa huee*, y es una palabra que usa la gente de aquellas partes <It is called by this name, *ixmatlatótotl*, because its song is like the way people speak. It says when it sings, *campa huee, campa huee*, and [that] is a word used by the people of these places>" (p. 695) (sound clip by Adam Dudley, ML432084251).

PŌPOCALES (Russet-naped Wood-Rail) "Tiene este nombre porque canta, diciendo *popocálex* a la puesta del Sol, y antes que sale canta, diciendo *popocálex*. <It has this name because it sings, saying *popocálex* at sunset, and before sunrise it sings, saying *popocálex*>" (p. 695) (sound clip by Jesse Lopez Herra, ML435824171).

QUILI-TON (parakeet). *Quili* may be imitative of the screechy calls of these parakeets, though this is nowhere explicit.

TACHITOHUIYA (Green Shrike-Vireo) "Llámase *tachitohuía* porque canta diciendo su cantar, es *tachitohuía* <It is called *tachitohuía* because it sings, saying its song, is *tachitohuía*>" (p. 708) (sound clip by Daniel Garrigues, ML437064221).

TAPAL-CATZOTZON-QUI (Barn Owl) "Es como la lechuza salvo que cuando canta soena como cuando golpean una teja con otra <It is like the Barn Owl except that when it sings it

sounds like when they hit one tile with another>" (p. 708). This is a nice description of the bill clacking so typical of the Barn Owl in flight.

TECOLŌ-TL (Great Horned Owl) "It has a deep voice when it hoots; it says, *tecolo, tecoolo, o, o*" (FC, p. 42). An excellent imitation of the typical calls of a Great Horned Owl.

TECUZIL-TŌTŌ-TL (Thicket Tinamou) "Y llámase así porque cuando canta dice *tecuh-cilton, tecuhcilton*. Tiene delgada la voz <And it is named this way because when it sings it says *tecuhcilton, tecuhcilton*. It has a thin voice>" (p. 695) (sound clip by Vicente Desjardins, ML435021671).

TEZOLOC-TLI (plovers/sandpipers) "Hacen ruido cuando vuelan <They make noise when they fly>" (p. 696).

TLATHUICICI-TLI (Canyon Wren) "Y su canto es *tlathuicicitli*. Canta en los tlapancos y sobre las paredes, y despierta la gente con su cantar. *Tlathuicicitli* quiere decir: 'Hola, hola, ya amanece!' <And its song is *tlathuicicitli*. It sings on the terraces and on the walls, and it wakes up the people with its singing. *Tlathuicicitli* means: 'Hello, hello, already it dawns!'>" (p. 709) (sound clip by Brayden Luikart, ML436865601.

APPENDIX 6

A Comparison of the Identifications of Martín del Campo (1940) with those of Hunn (2023)

If the species named is the same but the common name has changed, Martín del Campo's original common name is quoted in parentheses. Numbers in the third column indicate the following: 1: Hunn agrees with Martín del Campo; 2: Hunn disagrees with Martín del Campo and offers an alternative identification; 3: Hunn suggests an identification while Martín del Campo defers; 4: Hunn defers while Martín del Campo offer a suggestion; and 5: neither Hunn nor Martín del Campo have any idea as to the bird's identity. The number of cases of each is tallied below.

Martín del Campo	Hunn	
Resplendent Quetzal	Resplendent Quetzal	1
Mountain ("Mexican") Trogon	Mountain Trogon	1
Roseate Spoonbill	Roseate Spoonbill	1
Lesson's Motmot	Lesson's Motmot	1
Montezuma Oropendola ("Troupial")	Montezuma Oropendola	1
Yellow-winged ("Mexican") Cacique	Yellow-winged Cacique	1
Agami Heron	Northern Jacana	2
Red-legged ("Blue") Honeycreeper	Red-legged Honeycreeper	1
Lovely Cotinga	Lovely Cotinga	1
Turquoise-browed Motmot	Turquoise-browed Motmot	1
Emerald Toucanet	Keel-billed Toucan	2
Squirrel Cuckoo	Squirrel Cuckoo	1
Blue Grosbeak	Blue Grosbeak	1
Yellow-headed ("Parrot") Amazon	Yellow-headed Amazon	1
Scarlet Macaw	Scarlet Macaw	1
White-fronted ("Parrot") Amazon	White-fronted Amazon	1
Aztec Parakeet	Orange-fronted Parakeet	2
Red-crowned ("Parrot") Amazon	Red-crowned Amazon	1

Martín del Campo	Hunn	
hummingbirds	hummingbirds	1
Broad-tailed Hummingbird	Garnet-throated Hummingbird	2
Costa's Hummingbird	[unidentified hummingbird]	4
Broad-billed Hummingbird/Berylline Hummingbird	Broad-billed Hummingbird	1
[unidentified hummingbird]	[unidentified hummingbird]	5
Rufous Hummingbird	Mexican Violetear	2
Bumblebee ("Heloise") Hummingbird	Violet-crowned Hummingbird	2
Allen's ("Allen") Hummingbird	[unidentified hummingbird]	4
Cinnamon ("Cinnamomeous") Hummingbird	Cinnamon Hummingbird	1
Dusky Hummingbird/Blue-throated Hummingbird	Plain-capped Starthroat/White-eared Hummingbird	2
Ruby-throated Hummingbird	Broad-tailed/Lucifer Hummingbird	2
[unidentified hummingbird]	[unidentified hummingbird]	5
Rose-breasted Grosbeak	Rose-throated Becard	2
Rallidae	Russet-naped Wood-Rail	3
0	Tinamou	3
0	Great Tinamou	3
duck	duck	1
goose	Fulvous Whistling-Duck	2
Mexican Duck	Mexican Duck/female ducks	1
Mallard	Muscovy Duck	2
Greater White-fronted Goose	Black-bellied Whistling-Duck	2
Sandhill Crane	Sandhill Crane	1
0	Royal & Elegant Terns	3
0	shorebirds	3
American White Pelican	American White Pelican	1
American Coot	Common Gallinule	2
American Coot	American Coot	1
Black-crowned Night-Heron	Green Heron	2
Wilson's Snipe	Wilson's Snipe	1
Red-necked ("Northern") Phalarope	Black-necked Stilt, Wilson's and Red-necked Phalaropes	2
Cliff Swallow	Cliff Swallow	1
Barn Swallow	Barn Swallow	1
Snowy Egret	Snowy Egret	1
Little Blue Heron	Great Blue Heron	2

Appendix 6: Comparing Martín del Campo's identifications with Hunn's

Martín del Campo	Hunn	
Ocellated Turkey	Wild Turkey	2
Anhinga ("Water-Turkey")	Neotropic Cormorant	2
Western Grebe	Clark's Grebe/Western Grebe	1
Black Skimmer	Black Tern	2
Wood Stork ("Wood Ibis")	Wood Stork	1
Purple Gallinule	Purple Gallinule	1
American Bittern	American Bittern	1
Black-bellied Plover	Aztec Rail	2
American Avocet	American Avocet	1
Green-winged ("Common") Teal	Green-winged Teal	1
Blue-winged Teal	Blue-winged Teal	1
Canvasback	Canvasback	1
Hooded Merganser	Wood Duck	2
Bufflehead	Bufflehead	1
Ruddy Duck	Ruddy Duck	1
Northern Pintail ("Pintail")	Northern Pintail	1
American Wigeon ("Baldpate")	American Wigeon	1
Eared Grebe	Eared Grebe	1
0	Lesser Scaup/Ring-necked Duck	3
Mallard	Gadwall	2
Cinnamon Teal	Cinnamon Teal	1
Belted Kingfisher	Belted Kingfisher	1
Northern Shoveler ("Shoveler")	Northern Shoveler	1
Black-crowned Night-Heron	Black-crowned Night-Heron	1
Franklin's Gull	Laughing Gull	2
Western Grebe	Least Bittern	2
0	eagle/hawk	3
Golden Eagle	Golden Eagle	1
Common Black Hawk ("Crab-Hawk")	Ornate Hawk-Eagle	2
0	White-tailed Hawk	3
0	Northern Potoo	3
Northern Harrier ("Marsh Hawk")	Northern Harrier	1
0	Crane Hawk	3
Golden Eagle	Golden Eagle	1
Osprey	Osprey	1
King Vulture	Crested Caracara	2
Laughing Falcon	Laughing Falcon	1
Black Vulture	Turkey Vulture/Black Vulture	1

Martín del Campo	Hunn	
0	Great Horned Owl	3
Burrowing Owl	Striped Owl/Short-eared Owl	2
Common Raven	Common Raven	1
Jabiru	White-faced Ibis	2
Franklin's Gull	Franklin's Gull	1
falcon/Prairie Falcon	falcon/Prairie Falcon	1
Northern Harrier	Northern Goshawk	2
0	Peregrine Falcon	3
American Kestrel	American Kestrel	1
0	Aplomado Falcon	3
0	Peregrine Falcon	3
0	Prairie Falcon	3
0	Bat Falcon	3
Common Nighthawk	Bat Falcon	2
Merlin	Merlin	1
Loggerhead Shrike	Loggerhead Shrike	1
Black-backed ("Bullock's") Oriole	Black-backed Oriole	1
Band-backed Wren	*Campylorhynchus* species	1
0	Green Shrike-Vireo	3
Golden-fronted Woodpecker	Golden-fronted Woodpecker	1
Mexican Whip-poor-will ("Whip-poor-will")	Mexican Whip-poor-will	1
Crested Guan	Crested Guan	1
Barn Owl	Barn Owl	1
Barn Owl	Barn Owl	1
0	Burrowing Owl	3
Canyon ("Brown") Towhee	Canyon Towhee	1
Wren [*Thryothorus* species]	Canyon Wren	2
Eastern Meadowlark ("Meadowlark")	Eastern Meadowlark	1
sparrow	sparrow species	1
Vermilion Flycatcher	Vermilion Flycatcher	1
Red Warbler	Red Warbler	1
House Finch	House Finch	1
Inca Dove	Inca Dove	1
Montezuma Quail	Montezuma Quail	1
Slender-billed Grackle	Slender-billed Grackle	1
Great-tailed ("Boat-tailed") Grackle	Great-tailed Grackle	1

Martín del Campo	Hunn	
Red-winged & Yellow-headed Blackbirds	Red-winged & Yellow-headed Blackbirds	1
Mourning Dove	Mourning Dove	1
Common Ground Dove ("Ground Dove")	Rock Pigeon	2
Curve-billed Thrasher	Curve-billed Thrasher	1
Northern Mockingbird	Northern Mockingbird	1
0	Lesser Goldfinch	3
Ladder-backed Woodpecker	Imperial Woodpecker	2
Common Chachalaca	West Mexican Chachalaca	2
domestic Wild Turkey	domestic Wild Turkey	1

Summary		
1. Hunn agrees with Martín del Campo	1	78
2. Hunn disagrees and offers an alternative	2	34
3. Hunn suggests, Martín del Campo defers	3	19
4. Hunn defers, Martín del Campo suggests	4	2
5. Neither of us has any idea	5	2
6. Martín del Campo suggested these	1 & 2 & 4	113
7. Hunn suggested these as alternatives	2 & 3	53
8. Hunn suggested these	1 & 2 & 3	131
9. Total species identified	1 & 2 & 3	133

Permissions

Permission for Use Has Been Obtained for the Following Figures

Figure 5. Fine Arts Museums of San Francisco.

Figure 6. Permission for Use (Reproduction) granted by Kunsthistorisches Museum, Wien.

Figure 7. Permission for Use granted by the Akademische Druck- und Verlagsanstalt, Graz, Austria.

Figure 25. Permission for Use granted by the American Ornithological Society.

Figure 36. Commercial use rights for Ehecatl purchased from Symbolikon.

Macaulay Library granted permission for the use of the images in Figures 8, 9, 10, 12, 13, 14, 15, 16, 17, 18, 19, 20, 21, 23, 24, 26, 27,28, 29, 30, 33, 35, 38. Macaulay granted permission for use of all sound clips in the Digital Collection. The photographers and audio recorders are identified in the figure titles or captions along with the Macaulay catalog numbers.

The School of American Research granted permission to use the English translations of the original Náhuatl text. These translations are quoted from the Fray Bernardino de Sahagún in C.E. Dibble & A.J.O. Anderson 1963. Copyright 1981. All rights reserved.

CC BY-SAs Apply to the Following Figures
Figure 1. Valley of Mexico CC BY-SA 3.0
Figure 3. Fray Bernardino de Sahagún CC BY-SA 4.0
Figure 39. CHIQUIMO-LI, Imperial Woodpecker (Geller-Grimm 2003 CC BY-SA 2.5) with historic range map (Clark and Brown 2020 CC BY-SA 4.0).

Public Domain
Figure 2. Moteuczoma Xocoyotzin

Eugene S. Hunn

References Cited

Alcántara-Salinas G. (2011). *Comparative study of Cuicatec and Zapotec ethno-ornithology, with reference to contextual variation in a time of environmental and social change in Oaxaca, Mexico.* [Unpublished doctoral dissertation]. University of Kent, UK.

Alcántara-Salinas, G., Hunn, E. S. & Rivera-Hernández, J. E. (2015). Avian biodiversity in two Zapotec communities in Oaxaca: The role of community-based conservation in San Miguel Tiltepec and San Juan Mixtepec. *Human Ecology, 43*(5), 735–748.

Anderson, E. N. (2011). Ethnobiology: Overview of a growing field. In E. N. Anderson, D. M. Pearsall, E. S. Hunn, & N. J. Turner (Eds.), *Ethnobiology* (pp. 1–14). Wiley-Blackwell, Hoboken, New Jersey.

Anderson, E. N., & Medina Tzuc, F. (2005). *Animals and the Maya in southeast Mexico.* University of Arizona Press, Tucson.

Atran, S. (1990). *The Cognitive foundations of natural history: Towards an anthropology of science.* Cambridge University Press, Cambridge, Massachusetts.

Berdan, F. F., & Anawalt, P. R. (Eds.). (1992). *The Codex Mendoza.* University of California Press, Berkeley.

Berlin, B. (1992). *Ethnobiological classification: Principles of categorization of plants and animals in traditional societies.* Princeton University Press, Princeton, New Jersey.

Berlin, B., & Kay, P. (1969). *Basic color terms: Their universality and evolution.* University of California Press, Berkeley.

Berrin, K. (Ed.). (1988). *Feathered serpents and flowering trees: Reconstructing the murals of Teotihuacan.* The Fine Arts Museum of San Francisco, California.

Coe, M. D. (1962). *Mexico.* Prager, New York.

Clark, K. B. & Brown, D. E. (2020). Imperial woodpecker (*Campephilus imperialis*). CC BY-SA 4.0 (https://creativecommons.org/licenses/by-sa/4.0/). In T. S. Schulenberg & B. K. Keeney (Eds.), Birds of the World. Version 2.0. Cornell Lab of Ornithology, Ithaca, NY. (https://doi.org/10.2173/bow.impwoo1.02).

Diamond, J. M. (1966). Zoological classification system of a primitive people. *Science, 151*(3714), 1102–1104.

Dibble, C. E., & Anderson, A. J. O. (Translators). (1963). *Florentine Codex, Book 11—Earthly things.* Monographs of the School of American Research and The Museum of New Mexico, Number 14, Part XII. The School of American Research and The University of Utah. Santa Fe, New Mexico.

Ellen, R. (1982). *Environment, subsistence and system: the ecology of small-scale social formations*. Cambridge University Press, Cambridge, UK.

Friedmann, H., Griscom, L., & Moore, R. T. (1950). Distributional check-list of the birds of Mexico, Part 1. *Pacific Coast Avifauna* No. 29.

García Quintana, J. G., & Lopéz Austin, A. (1988). *Historia general de las cosas de Nueva España, Fray Bernardino de Sahagún*. Alianza Editorial Mexicana, Cuaderna Nacional para la Cultura y las Artes, México, D.F.

Geller-Grimm, Fritz. 2003 Kaiserspecht, (*Campephilus imperialis*), male and female from Mexico, Museum Wiesbaden, MWNH, Naturhistorische Landessammlung. CC BY-SA 2.5 (https://creativecommons.org/licenses/by-sa/2.5/).

Haemig, P. D. (1978). Aztec emperor Auitzotl and the Great-tailed Grackle. *Biotropica, 10*, 11–17.

Howell, S. N. G., & Webb, S. (1995). *A guide to the birds of Mexico and northern Central America*. Oxford University Press, New York.

Hunn, E. S. (1977). *Tzeltal folk zoology: The classification of discontinuities in nature*. Academic Press, New York.

Hunn, E. S. (2008). *A Zapotec natural history: Trees, herbs, and flowers, birds, beasts, and bugs in the life of San Juan Gbëe*. University of Arizona Press, Tucson.

Hunn, E. S., and Brown, C. H. (2011). Linguistic ethnobiology. In E. N. Anderson, D. M. Pearsall, E. S. Hunn, & N. J. Turner (Eds.), *Ethnobiology* (pp. 319–333). John Wiley & Sons, Hoboken, New Jersey.

Johannes, R. E. (1981). *Words of the lagoon*. University of California Press, Berkeley

Karttunen, F. (1983). *An analytical dictionary of Nahuatl*. University of Oklahoma, Norman.

Lammertink, M., Gallagher, T. W., Rosenberg, K. V., Fitzpatrick, J. W., Liner, E., Rojas-Tomé, J., & Escalante, P. (2011). Film documentation of the probably extinct Imperial Woodpecker (*Campephilus imperialis*). *The Auk, 128*(4), 671–677.

Leon-Portilla, M. (1992). *The broken spears: The Aztec account of the conquest of Mexico*. Revised edition. Beacon Press, Boston.

LeRoi, A. M. (2014). *The lagoon: How Aristotle invented science*. Penguin, New York.

Majnep, I. S., & Bulmer, R. N. H. (1977). *Birds of my Kalam country*. University of Aukland Press, Auckland, NZ.

Martín del Campo, R. (1940). Ensayo de interpretación del libro undecimo de la <u>Historia general de las cosas de Nueva España</u> de Fray Bernardino de Sahagún – 11 las Aves (1). *Anales del Instituto de Biología, Tomo XI, Numero 1*. Mexico, D.F.

Miller, A. H., Friedmann, H., Griscom, L., & Moore, R. T. (1957). Distributional check-list of the birds of Mexico, Part 2. *Pacific Coast Avifauna* No. 33.

Minnis, P. E., Whelan, M. E., Kelley, J. H., & Stewart, J. D. (1993). Prehistoric macaw breeding in the North American Southwest. *American Antiquity, 58*(2), 270–276.

Molina, Fray Alonso de. (1571). *Vocabulario en lengua castellana y mexicana y mexicana y castellana*. Porrua, Mexico City.

Morales Vera, T. E., & Edmundo, T. (2006). Las aves de los *Comcáac* ["The birds of the Comcáac"] (Sonora, México). [Unpublished thesis.] Licenciado en Biología, Universidad Veracruzana, Xalapa, Veracruz.

Rea, A. M. (2007). *Wings in the desert: A folk ornithology of the Northern Pimans*. University of Arizona Press, Tucson.

Schoenhals, L. C. (1988). *A Spanish-English glossary of Mexican flora and fauna*. Summer Institute of Linguistics, México, D.F.

Sibley, D. A. (2014). *The Sibley guide to birds* (2nd ed.). Alfred A. Knopf, New York.

Terraciano, K. (2019). Introduction. An encyclopedia of Nahua culture: context and content. In J. F. Peterson, & K. Terraciano (Eds.), *The Florentine codex. An encyclopedia of the Nahua world in sixteenth-century Mexico* (pp. x–18).

Todorov, T. (1984). *The conquest of America*. Translated from the French by R. Howard. Harper & Row, New York.

Wikipedia. (2022). The lake system within the Valley of Mexico at the time of the Spanish Conquest around 1519 (https://en.wikipedia.org/wiki/Valley_of_Mexico).

Wikipedia. (2022). Moteuczoma Xocoyotzin, a late 17th-century portrait attributed to Antonio Rodríguez (https://en.wikipedia.org/wiki/Moctezuma_II).

Wikipedia. (2022). Fray Bernardino de Sahagún (https://en.wikipedia.org/wiki/Bernardino de Sahagún; from Museo Nacional de Historia, Mexico).

Wikipedia. (2022). Founding of Tenochtitlán (Codex Mendoza, folio 61r, Berdan & Anawalt 1992:127; with permission of the University of California Press).

Wikipedia. (2022). Moteuczoma's headdress (with permission of the Welt Museum Wien, Vienna, Austria) with anonymous text.

Wood, S. (Ed.). (2000–2020). *Online Nahuatl dictionary*. Wired Humanities Projects, College of Education, University of Oregon, Eugene.

Index of English and Latin Bird Names

Index of Náhuatl Bird Names

Eugene S. Hunn

Eugene S. Hunn

Subject Index